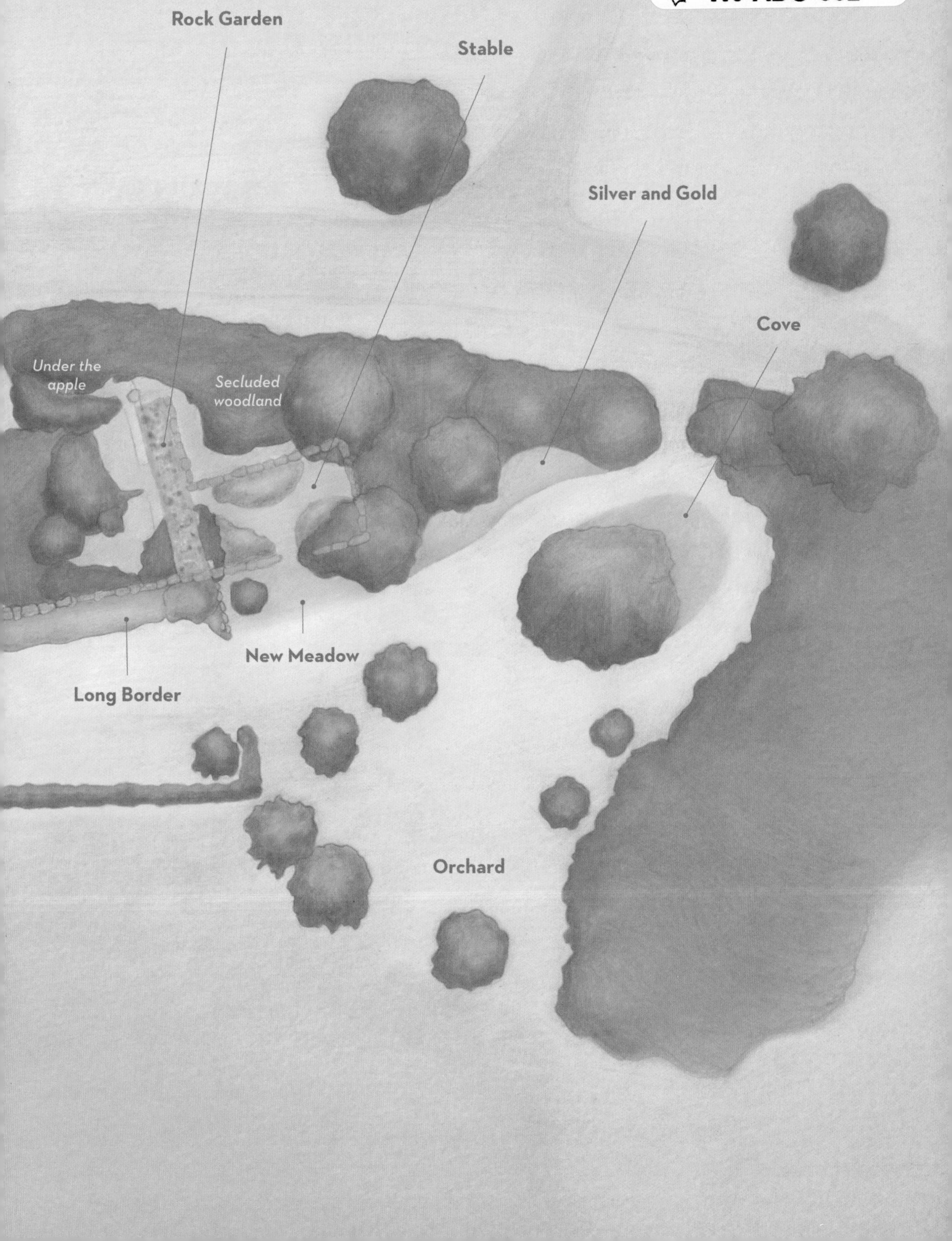

Rock Garden

Stable

Silver and Gold

Cove

Under the apple

Secluded woodland

Long Border

New Meadow

Orchard

SPIRIT OF PLACE

SPIRIT OF
PLACE

THE MAKING OF A NEW ENGLAND

GARDEN

BILL NOBLE

TIMBER PRESS · PORTLAND, OREGON

Published in 2020 by Timber Press, Inc.
The Haseltine Building
133 S.W. Second Avenue, Suite 450
Portland, Oregon 97204-3527
timberpress.com

Printed in China
Second printing 2020

Text and cover design by Stacy Wakefield Forte
Endpaper illustration by Anna Eshelman

ISBN 978-1-60469-850-3

Catalog records for this book are available from the
Library of Congress and the British Library.

FOR JIM AND SUE,

WHO MAKE IT POSSIBLE

CONTENTS

groundwork

I'm going to tell you a story of the pleasures and challenges, both aesthetic and practical, of creating a garden that feels genuinely rooted to its place.

Below the kitchen and dining room is a mixed planting of shrubs and perennials.

CLOCKWISE FROM TOP LEFT Fields slope away from the house toward the Connecticut River Valley and foothills of the White Mountains.

From the field below, a row of poplars at the garden's edge appear as sentries.

Built in the 1830s, our house is early Greek Revival and features old-fashioned shrubs on its east side.

I will also share the evolution of that garden, growing out of both its particular location and setting, as well as from my own interests and passions.

A sign at Dan & Whit's, the local general store in Norwich, Vermont, led us to the place we've called home since 1991. For two years, my partner James Tatum and I had been looking to buy a house, and seen dozens. Jim was intent on finding a contemporary house with a view; I wanted a farmhouse where I could garden. Every newer house, no matter how magnificent the view, seemed poorly put together, and the older homes were either situated in dark hollows or lacked a room large enough to accommodate Jim's Steinway Model B grand piano. A friend alerted us to an old farmhouse in Norwich that had land to garden—plus a view and room for a piano. The price had just been reduced to the upper end of our range. When another friend heard it was the house on

Bragg Hill that had belonged to Betty McKenzie, she exclaimed two words of advice: "Buy it!"

We made our way up the two steep miles of Bragg Hill to take a look. What we found was an early Greek Revival cape set close to the road, and it was the first house either of us cared for. We peered through the windows and could see it had high ceilings and a parlor that suited the piano. Jim led me through a passageway to show off the view, and my heart sank. How was I to build a garden to match that view?

A long panorama of foothills of the White Mountains unfolded to the east, across the Connecticut River Valley, and a classic Vermont farm hillside to the west. The land sloped away from the house, and there were tens of acres of lush green fields bordered by stone walls, with forest beyond. A hundred yards to

The Flower
Garden's generous
style and the way
it frames the view
complements
the scale of the
landscape.

the west, past overgrown piles of rubble, was a neighboring three-story house. My first impression was that to make a garden set against such a vast, magnificent landscape would require more effort than I had in me. But I also understood then that this would be the place we would call home, and I would have to confront its challenges.

As a self-taught gardener, I didn't have the advantage of a course of study in design or horticulture. I was fortunate in that my gardening life began in the gardens of artists who were primarily interested in creating outdoor places in which to live and work, where they could be close to nature and the elements, atmosphere and light, with a fullness of plants. For me, gardening is a way to experience nature and beauty in close engagement. In keeping with the spirit of that, I've made a garden here with perennial and shrub borders, foliage borders, a vegetable garden and orchard, a rock garden, meadow plantings, and where I continue to experiment with plants and fresh ideas.

The gardens where my practice began are located in nearby Cornish, New Hampshire. The artists who made them were amateur gardeners in the true sense, and used plants to create places filled with virtuosity and emotion. I learned that some of the greatest satisfactions of gardening can come from the process itself—from working the soil, observing how plants respond to their conditions, developing and refining design ideas, making changes, always striving for a more potent effect.

In the footprint of the former stable, low-growing alpine and woodland shrubs mirror vegetation found at higher elevations on distant ridge tops. OPPOSITE Rhythm and harmony in practice within the Flower Garden.

people and gardens

A garden can be an experience akin to being in the presence of a work of art, and while there can be transcendent moments in one's own garden, they are few and fleeting. Much of what gardening is about is the feeling of being connected to a place, fostering a sense of belonging, and becoming familiar with the natural rhythms and cycles of a particular piece of the earth. The intellectual and emotional aspects of gardenmaking can't be separated: cultivating plants, learning how they grow; observing their beauty, their variety of form and color; their hardiness, habit, and the qualities that make them good garden plants; whether they are workhorses or specialty plants—these are all as

CLOCKWISE FROM TOP LEFT A row of Lombardy poplars mediates the transition from garden to landscape along the edge of the field.

'King Alfred' daffodils are planted in drifts under the apples for a month's worth of flowering beyond the remaining barn.

Maple, apple, and crabapple trees trace shadows on the lawn, welcoming the start of a warm summer day.

rewarding as any physical sensation one might experience in a garden. These interests, the design of a garden and its emotional impact, as well as the plantsmanship involved in tending it, are what keep me engaged in my own garden.

I have found a great deal of inspiration in the artistry and experimentation of others—much of my own gardenmaking is a response to what I have observed in other gardens or learned from other gardeners. It is a distillation of design ideas and plants from dozens of places, many that I have visited, some that I have worked in, and a few I've only studied in photos or drawings. Part of my motivation for working in the field of garden preservation is that I believe that, in visiting gardens made by outstanding gardeners, we become better gardeners ourselves. I believe some of these spaces should be set aside for the public, so gardeners of all abilities can see them, learn from them, and be inspired by them.

The solitary aspect of gardening suits me, but so does the social. I delight in walking through my garden with one person or with a group of people. The conversations we have may be about plants, or about shared feelings and observations that arise in different parts of the garden, or about nothing at all. I enjoy seeing people explore the garden, especially when they circle back to take

Century-old apple trees are part of the farm's legacy.

OPPOSITE Stalwarts of the old-fashioned garden: bearded and Siberian irises, rose campion, roses, coral bells, and hardy geraniums.

a second look or simply stop and linger. I am curious about what people find of interest—it often means they see something I hadn't noticed before.

Our garden is also a place for entertaining. Commanding the north side of the house is a porch and deck with sweeping views of surrounding hillsides, and of the flower and vegetable gardens at its feet. Some visitors linger a long time on the deck before plunging in, drink in hand. From the deck, one feels closer to the sky and the elements. The garden is large and varied enough so people somehow seem to disappear into it in solitary wanderings, intimate conversation, or voluble gatherings. My pleasure in these gatherings is that friends are able to experience natural beauty in the company of others.

Early on, I learned to appreciate gardening in places with a history. I enjoy working within a framework created by others, enhancing it, and playing off it,

and this garden also says something about preservation. Its site was farmed for nearly 200 years, much of that time as a dairy farm surrounded by acres of cultivated fields and woodlots. The farmstead, with its 1830s Greek Revival house, barns, and outbuildings, was cultivated by the McKenzie family for most of the 20th century. The present site continues many of the beds they laid out, as well as trees and shrubs they planted over those decades.

Mine is an amateur's garden, begun shortly after I first came to professional gardening. We moved to Bragg Hill in the fall of 1991, when I was still employed in my first job as a gardener, with no formal training and only a few years of relevant experience. In this way, my garden's development tracks the evolution of my career in the world of public gardens and preservation. It is

FAR LEFT At the north end of the Rock Garden, where the foundations are less secure, *Juniperus communis* 'Effusa', phlox, sedum, and thyme cover the ground. Pasque flower's airy seed heads are its second season of interest.

MIDDLE These plants had been established years earlier. After an initial pruning and cleanup, they were ready to put on a show again.

LEFT A cool, moist spot by a bench is filled with arctic willows, heathers, false Solomon's seal, umbrella leaf, and dwarf *Darmera peltata*.

autobiographical, reflecting my trajectory in this occupation—from market gardener to garden preservationist to garden designer. Key parts of it were born out of working at other sites. I made the Flower Garden, for example, while I was working in gardens in Cornish. The Rock Garden started when I was restoring its counterpart at The Fells, and other parts evolved as I was exposed to the larger world of gardens.

In this sense, my garden expresses a range of interests and experiences. It is a place where I have experimented with plants and developed ideas that I've applied to other gardens as part of my design practice. It is a kind of laboratory or workshop, as well as a repository for plants that once had a place in horticulture, but are no longer propagated or widely available today.

germination

I did not grow up gardening, nor did it come naturally to me—I was in my 30s and had already started down a couple of career paths before I found what a pleasure it could be to work with plants and in the soil. I had grown up on a suburban lot on the edge of a wooded hillside in Connecticut, where I traced old cow paths and built forts, and later on escaped to the woods for teenage mischief. Something must have taken root, for later on I took terms off from college to hike in the Sierra Nevada, and during graduate school in Toronto, I lost the urge to commute into the city, opting instead for a rented farm in the country. But through those years, a specific interest in plants and horticulture eluded me. I ran in the woods, hiked mountain trails, and worked summers on the farm more for the experience of being outside in nature than for any particular interest.

It was only after moving to Vermont in the 1970s and trying my hand at different jobs that I came to gardening. I don't regret this lack of clear career path. The experiences of my 20s and early 30s, working construction and cabinetmaking, community organizing and farming, provided the background for the work that I ultimately turned to—working with gardeners and community activists who are passionate about saving gardens, and sometimes restoring them to share them with the public.

My entry into gardening was as a market gardener. By a rather circuitous route, I found myself farming 10 acres of vegetables, first in Cornish, New Hampshire, and then across the Connecticut River in Windsor, Vermont. What brought me to farming, though it was a path I had no idea I would embark on, was an irresistible urge to work outside with my hands.

After college, I had enrolled in a graduate program in medieval studies at the University of Toronto. One of the aspects of the Middle Ages that most intrigued me was the connection between monastic spirituality and rural life. I didn't last long in academia, or in Toronto, and soon ended up living in rural Ontario.

When my time in Canada came to a close in 1976, I made my way to Northern New England with a friend, where we both found jobs working

Against convention, I've placed my Vegetable Garden front and center. Perennial herbs are planted along the front, raised beds with rotating crops in the middle, and permanent rows of raspberries and asparagus behind the birdbath.

construction. Eventually I found more rewarding work as a community orga-
nizer in the New Hampshire mill town of Lebanon, where I helped establish
community gardens and a food co-op, along with other neighborhood self-help
programs.

By the early 1980s, I'd grown eager to find a way to make a living that
allowed me to be outside and work with the soil. I found land to rent, bought
a tractor, and planted 10 acres with a mix of corn, tomatoes, salad, and vine
crops for sale at a farm stand and farmers' market. Satisfaction came from
growing good quality produce and marketing it directly to my customers. It
was hands-on at each step, from direct seeding salad crops at 10-day intervals, to cutting and crating lettuce the morning of a market to staying until it
was all sold. I enjoyed using the community organizing skills I had acquired
in collaborating with local growers to start the Cornish Farmers' Market.
Although farming is long in the past, I still take enormous pleasure in seeding
salad crops as soon as the soil can be worked in April and bringing something
in from the garden to eat until late in fall.

Vegetable farming was a satisfying life, but an exhausting one, and something I couldn't sustain. I was grateful for the $5,000 I netted my first year, but
a few years later I sold the tractor when I realized my income had been exactly
the same. I began to search for a way to incorporate some of my other skills and
ideas with gardening. I wanted to combine my interest in history with horticulture, preferably in gardens accessible to the public. By luck, I found a tempo-
rary position working in the gardens of sculptor Augustus Saint-Gaudens at
Aspet, his house and studio, also located in Cornish. This grew into a full-time
job restoring and maintaining the historic gardens at the Saint-Gaudens
National Historic Site, and eventually led me to become involved with a number of other Cornish gardens designed and made by artists.

planting history

In the 1880s and 1890s, about a hundred artists, writers, and other political
and cultural figures were drawn to the natural beauty of the Connecticut River
Valley, and gathered around Saint-Gaudens, the foremost American sculptor
of his day, in what came to be known as the Cornish Artists' Colony. Many had
traveled in Italy and France, and for them the views across the river to Mount
Ascutney in Vermont recalled the landscape of Tuscany. Through the efforts
of Charles Platt, Ellen Biddle Shipman, and talented artist gardeners such as

Stephen and Maxfield Parrish and Saint-Gaudens himself, a new classical style of American garden took root.

Platt, a painter and architect, was a key figure in this development. In 1892, he traveled to Italy to study Renaissance garden design and soon published *Italian Gardens*, the first book of its kind written in English. His own garden in Cornish was informed by Italian ideas about the integration of house with garden, and garden with surrounding landscape; this led to his first major commission as an architect, the design of an Italian villa on a neighboring hillside. He soon brought this practice of designing houses and gardens as a unified whole into some of the wealthiest precincts in the country.

Platt's neighbor and colleague Ellen Biddle Shipman, named "Dean of American Women Landscape Architects" by *House and Garden* in 1933, also developed a national roster of clients. She trained with Platt and collaborated with him on dozens of residential gardens, becoming much sought after for her lush planting design. She was an expert plantswoman, experimenting with horticulture and planting combinations in her own garden in neighboring Plainfield. Long involved with Aspet, Shipman's knowledge and keen sense of design were crucial to the garden's longevity.

As gardener at Aspet, my initial assignment was to restore a half-mile of hundred-year-old white pine and hemlock hedges. This led to the restoration of the formal garden originally planted by Saint-Gaudens and later reworked by Shipman. I learned on the job as well as in workshops and training programs sponsored by the National Park Service, owner of the Saint-Gaudens site. I took classes in plant identification, horticulture, and garden history and preservation, and eventually earned a certificate in gardening arts from the Arnold Arboretum of Harvard University. I also worked with the current owners of Cornish Colony gardens to help make them livable, maintainable gardens rooted in the past.

This was the start of my lifelong involvement with garden history and preservation. Throughout my career, I felt somewhat at a disadvantage for not having studied in a traditional horticulture or landscape architecture program, but the knowledge and skills I acquired working in this garden and the formative experiences of restoring other gardens gave me sufficient foundation.

My desire to expand my horticultural skills led me to The Fells, the New Hampshire garden of Clarence and Alice Hay, an early preservation project of the Garden Conservancy. The five years I spent there as director of landscapes enriched my professional vocabulary and solidified an interest in gardens that are in equal measure about ecology and beauty.

CHARLES A PLATT *His Place at* CORNISH

The Connecticut River Valley and the
CORNISH ARTISTS' COLONY

CLOCKWISE FROM TOP LEFT View over the Val di Sieve in Tuscany. • The Connecticut River Valley and Mount Ascutney in the 1880s. • Platt's integration of house and garden set the tone for a Cornish-based style that merged Italian concepts with vernacular form. • Garden, architecture, and established trees work in harmony in the Platt garden. • Like Saint-Gaudens, nearly all artists enhanced their view of Mount Ascutney. • The painter Barry Faulkner wrote: "The landscape of Cornish was a revelation of delight: the down-pouring of the hills into the rich meadows by the Connecticut River, the domination of Mount Ascutney, towering over land and river; the size and density of oak, maple, and birch, nourished by heavy river fogs and the grand bulk of the giant white pine, all these filled me with wonder."

The Connecticut River Valley was first settled in the 1760s, and by the 1880s was mostly open farmland, with clapboarded farmhouses surrounded by hundred-year-old apple orchards. The landscape architect and garden book author Rose Nichols said of Cornish, New Hampshire, "few parts of New England bear so strong a resemblance to an Italian landscape as the hills rising about the banks of the Connecticut River opposite the peaks of Mount Ascutney."

The region attracted artists from New York wishing to escape the heat and noise of the city to pursue their work. The Cornish Colony formed in 1885 when Saint-Gaudens rented a house and studio, and was soon joined by other sculptors, painters, and writers. Daytime was for work in their studios, while afternoons and evenings were given over to recreational and social activities. Gardening soon became a shared passion for summer and year-round residents alike.

Acting as their own designers, these artists developed a new style of American garden, inspired by the ideals of Italian Renaissance garden design, and incorporating local materials and craftsmanship and old-fashioned hardy plants. Platt, Shipman, and Nichols were three prominent practitioners among them. Artist Stephen Parrish labored for 10 years making Northcote, which was widely regarded as the best gardened of them all; his son Maxfield Parrish built The Oaks on a neighboring hillside. These artists' gardens exhibited great originality and creativity, yet they also shared common characteristics, noted by design journals of the day.

"Nearly every garden enhanced its view of Mount Ascutney," wrote Mary Adams French, wife of sculptor Daniel Chester French. The Saint-Gaudens family enjoyed their meals while gazing at Mount Ascutney, "just as in Sicily they look toward Etna and in Japan toward Fujiyama," she said.

There was a close connection of house, gardens, and landscape, and the fitting of the house to the site. A series of garden rooms connected by sightlines and straight paths was often on axis with the house. A strong connection was made between indoors and out.

There was often a covered place to linger outdoors: a piazza, porch, or loggia. The gardens usually featured a central pool, fountain, sundial, or other special feature, often aligned with the house. French noted the artistic use of garden ornament in Cornish gardens in her book *Memories of a Sculptor's Wife*: "those little touches which no one but an artist would have thought of perpetrating. . . . A few columns, a stone floor against the house, and an amphora or a colored relief. . . . One might have been in Italy."

These gardens typically took advantage of established trees, especially century-old apple trees. They made good use of vernacular materials and native plants, such as white pine and hemlock for hedges, and local granite and Vermont marble.

Hardy, herbaceous perennials were favorites, many of them old-fashioned plants popular in Colonial Revival gardens.

Above all, the charm of Cornish gardens lay in the way they blended informal plantings with formal design principles; the formal and informal were contrasted and meant to complement one another. As they were gardens of artists, they were often composed as a painter might compose a painting.

CLOCKWISE FROM TOP LEFT At The Oaks, Maxfield Parrish aligned a reflecting pool with the house and loggia, reminiscent of his tour of Italian gardens. • Anne Parrish surrounded by hollyhocks and phlox, the mainstays of the old-fashioned garden at Northcote. • "Lombardy poplars have more than once been used with excellent effect by Cornish gardeners, and, what is rare, with reserve."—writer Frances Duncan • The Oaks composed as an artist might make a painting. • Native white pine and hemlock shaped into impressive hedges at Aspet.

The Fells brought me into the orbit of some extraordinary gardeners, thanks to Frank Cabot's vision of a national garden preservation organization. For the next 15 years, as director of preservation for the Garden Conservancy, I had the privilege of working with a host of talented and innovative gardeners. Alongside Antonia Adezio, the Conservancy's executive director and president, I collaborated with artist-gardeners responsible for some of the most exceptional gardens of our time, as well as many practitioners who have rescued, restored, and skillfully maintained other significant sites. Since leaving the Conservancy, I have continued to work with gardens that are in transition from private to public, and I've developed a practice in garden design and restoration. I look for opportunities to revitalize existing gardens, as well as create new ones of lasting quality.

learning in artists' gardens

Augustus Saint-Gaudens approached the making of his garden the way a sculptor would: he made quick sketches on paper to think through the three-dimensional aspect of what he wanted to build, or he set up a maquette to study the location of a feature—a pergola, for example. He had assistants dig hundreds of small white pines out of the fields to plant into tightly trimmed hedges. Eventually, a half-mile's worth of hedges formed a series of garden rooms. One included a marble fountain outfitted with sculpture gilded in his workshop, to be viewed from a wooden bench, with Mount Ascutney in the distance. In the hedged terraces off the house, he arranged flowerbeds of colorful perennials, annuals, and bulbs, ornamented with more gilded figures and a semicircular bench.

Saint-Gaudens's use of traditional building crafts and local materials with garden furnishings he designed were hallmarks of the Arts and Crafts approach to gardenmaking. Lombardy poplars at the four corners of the house and studios, ornamented with plaster casts from the Parthenon, evoked memories of Italy.

Over the decades, Saint-Gaudens's original design gradually lost some of its luster. The hedges had grown too tall and wide; the Lombardy poplars were gone; the succession of perennial bloom in the flower garden fell flat after June. Apple trees, lilacs, and vines were in need of taming, and the lawns were

Saint-Gaudens approached gardenmaking the way he approached sculpture: he set up a maquette to study the location of a pergola and laid lath on the lawn to determine the shape of new garden beds.

worn threadbare. I spent my initial summer tying up limbs and hand-pruning hundreds of pines and hemlocks. This combination of renovation and improved maintenance soon began to show results. The idea that I could rejuvenate a historic garden and see results was eye-opening for me.

Renovating the formal garden was a more complex affair. The flower garden had been redesigned and replanted over the course of a century. A team of historic landscape architects became involved, and studies were done to understand each successive period. Although I was a part of this process, I had already begun to make improvements while planning inched forward. I had a hunch that the ultimate solution would be to restore the garden as reconceived by Ellen Shipman in the 1940s. The footprint of the garden was almost entirely a result of that makeover, and the plantings did, in fact, bear some resemblance to a detailed planting plan of Shipman's design.

I renovated the formal garden, working intensively over a three- to four-year period, removing aggressive plants and tracking down desirable ones to reintroduce. For reference, I worked from Shipman's landscape plans, historic black-and-white photographs, and a description of the garden she published in the *Bulletin of the Garden Club of America*, as well as Saint-Gaudens's notes, handwritten in his copy of the 1902 book *A Woman's Hardy Garden*, by Helena Rutherfurd Ely. Shipman's notes on her plan about maintenance helped recreate a garden that had historic precedent, but at the same time felt fresh and invigorating.

The beds at Aspet began to flourish. Flowers long absent, such as delphiniums, regal lilies, hollyhocks, asters, and old-fashioned annuals now enlivened the scene. Perennials with lush foliage and colorful flowers bloomed May to October. I was grateful to have the architectural framework of the hedges Saint-Gaudens planted to work with, and the garden took on the feel of belonging to an actual gardener—that it was a personal garden, not a historic site bound by institution.

Working in this garden influenced how I would make my Flower Garden at Bragg Hill, especially as I became acquainted with Shipman's style of closely packed perennials, focus on plant combinations, and finely orchestrated sequence of bloom. Although her typical borders were high maintenance and required skilled gardeners and frequent swapping out of annuals, the borders she redesigned at Aspet were somewhat less demanding and introduced me to a palette of successful perennial garden plants.

branching out

As the owners of neighboring Cornish Colony gardens saw the garden at Aspet come alive, some asked me to work with their gardens too. Beginning in the late 1980s, I helped reinvigorate other gardens associated with Platt, Shipman, and Stephen Parrish, including Parrish's own garden, Northcote. Parrish sculpted the hillside, built stone walls, wooden balustrades, and pergolas, and laid out colorful beds with a stunning array of trees, shrubs, conifers, roses, perennials, annuals, and bulbs. Reading his garden journals, inspecting his plant lists, and studying photographs of Northcote taken by his son Maxfield not only helped me restore parts of the garden, but also had an indelible effect on my own. I learned how to assemble a garden out of a wide variety of herbaceous and woody plants, some of which provide year-round height, volume, and structure, while others take center stage at their season of interest. I'm not ashamed that some of the planting ideas in my garden are direct quotes from Northcote.

At The Fells, I led the restoration of a garden whose inspiration was more about a love for the rugged New Hampshire landscape than an artist's memories of Italy. It was also a plantsman's garden, with a legacy of choice alpine and rock garden plants. Originally a summer cottage for John Hay—one of Abraham Lincoln's secretaries and later secretary of state—The Fells became an estate garden that acted as an encyclopedic record of horticultural study over 60 years of stewardship under Hay's son Clarence and his wife, Alice. Clarence Hay loved the rocky, acid-soil pastures bordering Lake Sunapee, and made liberal use of the local granite and native plants to create formal walled gardens, perennial and rose gardens, rhododendron dells, and an extraordinarily beautiful rock garden that seemed to grow organically out of the granite-covered hillside.

The landscape had been well maintained during Clarence Hay's lifetime, but endured a period of neglect after his death in 1969. Eventually the property was transferred to the U.S. Fish and Wildlife Service, and the Garden Conservancy began rehabilitation in 1993. First, we recruited volunteers from the community to help with the hard work of reversing that neglect. We began by removing decades of overgrown trees and weeds along an entry drive bordered by masses of rhododendrons, native groundcovers, and ferns, and then gently pruned the majestic trees surrounding a walled woodland garden that retained the romantic spirit of a garden ruin.

Clearing away that overgrowth revealed a hidden gem, and as rewarding as that was, it was only the beginning—next and most noteworthy, we re-established a 100-foot-long perennial border. My reworking of the border reflected the tastes and plants of the Hays' long-gone design; volunteers planted the garden in May, and we celebrated its rebirth with a party that July. The public began to take notice.

Most challenging of all was the restoration of the ¾-acre rock garden that had been 10 years in the making. Hay laid out its granite boulders and stepping stones in the 1920s and 1930s, on a hillside that sloped away from the house toward the shore of Lake Sunapee. A rill emerged at the top of the garden and water trickled through the length of it, resting in pools and spilling over cascades, until finally disappearing into the woods. It was an entirely believable alpine scene crafted out of raw pastureland. Taking advantage of the existing glacial erratics and topography, Hay created a number of micro-environments and filled them with selections of alpine and rock garden plants, each matched to their cultural requirements.

This project taught me there can be various approaches to renovating a rock garden: one can either scrape everything away and start over, or identify long-term survivors that are salvageable and work with those. We carefully weeded and rejuvenated some sections and started fresh in others.

At The Fells, I saw that garden preservation did not merely need to attempt to repeat what had gone before—it could be dynamic, leading to gardens that bridged the decades between an initial design and one that was authentic, beautiful, inspiring, and educational for people today. I also learned how powerful a group of committed volunteers can be, both in hands-on work and as advocates for rescuing a garden and making it an essential part of a community.

INSPIRATION ABROAD

I met Jim just as I was transitioning out of vegetable farming, and his understanding and support for me while I figured out what came next was a gift. A professor at Dartmouth College, his teaching schedule and fellowship awards have facilitated international travel, and that has also been to my great advantage.

For a period, we enjoyed a series of autumn stays in Italy. I sought out gardens on these trips, many of them familiar to me from Platt's *Italian Gardens*. Gaining access to those villas often involved writing letters to the current owners and asking their permission to visit and photograph them. I saw firsthand some of the basic principles of the Italian Renaissance garden Platt points out in his book: the harmony of arrangement between house, garden, and landscape; and how a garden, with its regular walks and terraces, can be part of a progression that leads to groves and the larger landscape beyond.

I was able to join Jim during fellowship stays at the Rockefeller Foundation's Villa Serbelloni on Lake Como, and the Ligurian Study Center in Bogliasco on the coast near Genoa. Spending time in each of these villas gave me a sense of how the Italian garden was conceived of as a place to conduct a civilized life. It was where one lived outside, meals taken on terraces, books read in shady nooks, and where lively conversation stimulated the mind on walks or in the presence of awe-inspiring scenery. I breathed deeply in both settings, and have brought some trace of them home in my memory and in my garden. I am told the garden became more Italian in feel after these trips—and I can see this in the way its poplars and conifers lead the eye toward the view, and in how the garden unfolds with a sequence of distinct experiences.

Trips to Rome took Jim and me to a number of Renaissance gardens in and around the city. We spent a November in a farmhouse just outside Florence, where we were able to visit the great Medici villas of Tuscany and the restored gardens of La Pietra and Villa Gamberaia. I was attracted to the history and details of these places, the layers of architecture and patina, centuries-old trees and hedges and the aura of romantic ruins. I got to know architects and landscape architects charged with maintaining and restoring monuments of Italian Renaissance garden history.

Italy showed me how to look at the garden as architecture, and how the garden could relate to a landscape not unlike that of my home in the Connecticut River Valley. Even so, I knew there were other well-known, more plant-focused gardens I needed to see and learn from elsewhere in Europe.

In England, I discovered how the climate and expanded palette of shrubs and perennials, coupled with stately brick walls and yew hedges, encouraged gardeners to take a more spirited, spontaneous approach to planting design. I discovered plants and planting combinations that would inform my own work. At Great Dixter in Sussex, I learned how to extend the growing season by interplanting bulbs among shrubs, and how to coax a longer season of bloom out of perennials. Repeated visits to the Beth Chatto Gardens and immersion in her books has taught me, as a plantsman, to pay equal attention to ecology and beauty in composing a garden.

Jim's participation in a conference in the Netherlands was the impetus to visit Hummelo, landscape designer Piet Oudolf's garden (now closed to the public), along with that of the modernist landscape architect Mien Ruys and a number of German botanic gardens. Karl Foerster's garden near Potsdam and the Sichtungsgarten Weihenstephan outside Munich brought me to the headwaters of the German ecological approach to gardenmaking. Foerster's garden was full of resilient and beautiful plants—many of them North American native perennials and grasses, some of which he hybridized to make them more adaptable to the severe continental climate and poor soils of the region. An October visit showcased these plants' autumnal beauty, and presented a palette a bit more suited to New England conditions than many English borders.

At Weihenstephan, I observed how horticulturist Richard Hansen had developed a new approach to planting design based upon a rigorous analysis of plants and their garden habitats, and how, by uniting ecology and design, he combined plants from similar habitats with an eye toward functionality and beauty. His influence extended to botanic gardens and parks, where I discovered exciting possibilities in working with plants from the steppe regions of Eurasia. Oudolf has carried the work of Hansen and others forward, and brought this combination of ecology, design, and functionality to a much wider public. Characterized by the juxtaposition of its inventive, sculptural framework and the wildness of both flowering and senescent plants, Oudolf's garden at Hummelo opened my eyes to the ephemeral beauty of steppe and prairie plants.

CLOSER TO HOME

The opportunity for eye-opening international trips like these came every couple of years. Most of my time spent studying and learning during this period was spent closer to home. A few public gardens influenced me enormously: the Arnold Arboretum and Garden in the Woods in Massachusetts were places to visit and get ideas; Montreal Botanic Garden introduced me to a northerly garden with plants more suited to Vermont.

Wave Hill in New York revealed the possibilities for reimagining an established landscape with a strong architectural design and thoughtfully placed new plantings as a public garden. Marco Polo Stufano had arrived there in 1967, when the former private estate in the Riverdale section of the Bronx became a New York City cultural institution. He found gardens in decline, but a site that had been well laid out, with stately trees and stunning views of the Hudson River and the Palisades. By the time I first saw Wave Hill, Marco had been there 20 years, and the gardens had attained a character unlike any I'd seen in my U.S. travels. The flower garden and wild garden made vivid impressions on me, as did the herb garden and alpine house. For years, I kept two color photographs pinned to a bulletin board above my desk: the flower garden with its mix of shrubs, roses and perennials, and color-themed sections; and the wild garden with its picturesque summerhouse and clouds of densely pruned yews presiding over a rumble of wildflowers and majestic views.

At Wave Hill, Marco, working with John Nally, and fellow gardeners proved that the Arts and Crafts garden approach was still valid for contemporary American gardens. What I saw there gave me greater confidence to experiment with horticulture. Wave Hill was full of fascinating ideas and plants that could be translated into gardenmaking in the more rigorous New England climate.

It was some time after my introduction to Wave Hill that I got to know Marco. He and I would go on to visit dozens of gardens and work with many of them on behalf of the Garden Conservancy. Marco's eye for quality and his ability to assess a garden was invaluable. Garden visits with him and the discussions and debates that followed helped fine-tune my own sense of the elements that make up a garden. Inevitably, that fine-tuning and an analytical approach have made my garden a better garden.

I have also made my garden in the context of regular exposure to extraordinary personal gardens and the people who made them, as you'll notice throughout this book. Contemporary gardeners such as John Fairey in Texas, for example, have helped me to understand how to use light, the greatest free commodity in a garden. From George Schoellkopf in Connecticut and Nancy Goodwin in North Carolina, I've learned how it's possible to compose scenes that feel entirely natural with rare plants, and from Ione Chase in Washington state, how plants that might be thought of as common can be used to create refined garden beauty. And it's from mentors like Frank Cabot and colleagues associated with the Garden Conservancy that I've come to appreciate the idea of the garden as both an emotional and aesthetic journey.

I could not have made my garden as it exists without the plants that propagators, plant explorers, and nurseries have brought to the market over the past 30 years. Much of my garden has been outfitted with plants from specialty nurseries: Plant Delights Nursery in North Carolina; Heronswood and Forestfarm in the Pacific Northwest; Broken Arrow Nursery and Opus in southern New England; and especially Cady's Falls Nursery and E.C. Browns' Nursery in Vermont.

The gardens that speak to me the most are those of extraordinary plantsmen and women who have grasped the raw material of a site, and, through their engagement with place, have created spaces that are both eloquent and useful. My own garden has been informed by these diverse places, some of them conceived in traditional styles and others that broke new ground in horticulture. What they have in common, as the poet Marianne Moore puts it, is "a place for the genuine."

WHAT MAKES A GARDEN?

At the Garden Conservancy, one of the primary aspects of my job was to discover what makes a given garden significant, and then to communicate that to others. One of the things I learned from my colleagues was what questions to pose when assessing that significance. Some of the questions I ask when looking at a garden are:

How does the garden relate to the larger landscape? How does it connect to the natural and cultural environment, and does it evoke a sense of place? How does it relate to the site's architecture? How would I describe its design, and what gives the garden its form and structure? How does the garden maker create space? Who made the garden? Is the garden a personal expression?

How did the raw material of the site suggest the outcome of the garden, and where did the garden maker find inspiration? How did it develop over time?

How does the gardener use plants and create compositions with them? What characterizes the quality of plants and their cultural conditions? Are there plant collections?

How is the garden maintained, and what is the quality of its maintenance? Is the garden durable? Does the garden contribute to the environment, to wildlife, biodiversity, and a healthy ecosystem? How is the garden used? Is it an environment to be lived in?

Above all, does the garden evoke an emotional response?

CLOCKWISE FROM TOP LEFT The gentle sound of murmuring poplar leaves carries into the garden. • *Rheum australe* brushes against foundation walls. • *Thuja occidentalis* 'Rosenthalii' with contrasting willows *Salix candida* 'Silver Fox' and *S. alba* var. *sericea*. • A bench sited to take in a view and pause for a moment of reflection. • Spirit of place: garden and landscape as one. • Spires of *Actaea simplex* 'White Pearl' open in mid-October to the relief of late-season bees and syrphid flies.

bragg hill beginnings

In addition to the house's charm and gorgeous views, the place on Bragg Hill had much to recommend it.

Massive granite retaining walls, twin sugar maples and century-old apple trees speak of the long history of the farm.

When we found it in 1991, the landscape had been shaped by close to 200 years of farming, especially by the previous owners, who had farmed and gardened the property for over 60 years. Massive, lichen-covered granite stone walls created a yard on the east front, watched over by two venerable sugar maples. Large lilacs and old-fashioned flowering shrubs anchored the house. A barn with weathered shingles and granite foundations stood against the road, while century-old apple trees and a vegetable garden bordered by the lush green grass of the fields lent an air of husbandry and cultivation.

After nearly losing the sale to competing buyers who were going to jack up the house and move it into the field away from the road, our offer was accepted, and we moved that fall.

unearthing a garden

In cleaning up the collapsed barns and decades-old plantings, I uncovered the history of the farm and its gardens, and the potential for a new garden with strong bones revealed itself. What emerged under our ownership was an effort to restore the best of the remnant garden and the creation of new spaces within the framework of the farm.

The evolution of this design, how I went about giving the garden form and structure with plants, is the heart of this book. My greatest challenge was to make a garden in harmony with the expansive landscape of the Connecticut River Valley surrounding it, as well as one that would be manageable in scope, yet diverse enough to be interesting. Early on, I established an overall approach for restoring walls, planting trees and hedges, creating new garden spaces and tying them together into a satisfying whole. Old flowerbeds became settings for new plantings. Once cleaned up, the ruins of a hay barn, dairy barn, and stable presented opportunities for new garden areas. The design that emerged was shaped not only by the place and its history, but also by my work with other gardens, both historic and contemporary.

Apple trees for cider and for keeping grow at the edges of fields.

The house faces south. Decades-old roses on pillars flank the front door.

OPPOSITE Our land was first settled in the 1760s, and was a working dairy farm into the 1950s.

THE HOUSE

Our house is set 25 feet from the road. Like most Vermont farmhouses of the 1830s, it was carefully sited and built close to the road to ease winter access. We came to appreciate other practical aspects of the house's siting—it faces south to southeast to capture winter sun, and an ell extension diverts winter wind out of the northwest so that we rarely feel a draft inside. We see the winter sun from the time it rises over the hills to the east until it sets over the neighboring ridge. The house was not sited for its view, and in fact it seems to turn its back on the view, but it was built at a time when access to the animals in the barn and a wood pile for heating were considered paramount.

The house is a solidly built, clapboarded early Greek Revival farmhouse with simple details. The main part of the house includes a kitchen and two parlors, with a birthing and dying room downstairs, and three bedrooms upstairs. The ell extension was most likely added later when the farm became more prosperous, providing a covered workspace and seasonal apartment for a hired hand.

The rooms are ample, and because of their height and the placement of windows, as well as the wide pine-board floors, the living rooms are illuminated with golden light. Most of the house retains the original plaster walls and ceilings, and vintage wallpaper in shimmering gold and cream lines the dining

room walls. The former owners appreciated the antiquity of the house and kept up with its maintenance, and their improvements didn't detract from its 19th-century charm and comfort. It is a very practical house.

THE LAND

The land on which our home sits was first settled in 1767 by Jonas and Hannah Richards, who, like many of the state's first settlers, came to Vermont from northeastern Connecticut. Located on the western banks of the Connecticut River, across from Hanover, New Hampshire, and Dartmouth College, the Vermont town of Norwich was first settled in 1763 and named after Norwich, Connecticut. It's part of a region known as the Upper Valley, where the economic, cultural, and social life of the two states is unified by the river.

Richards would have first accessed his new lot in Norwich by walking northward along the frozen banks of the Connecticut River in late winter, before the ice was gone. He spent his first summers felling trees on the 100-acre subsistence farmstead, and built a log cabin where his first son was born in November 1767. Early Vermonters made farms by clearing densely wooded hillsides, as these sites proved more habitable than the area's marshy lowlands. At the elevation of 1,150 feet, the Richards farm looked out over other settlements then being cleared out of the woods. By the 1830s, the farm prospered, and the present house was built.

By the mid-1800s, the farm had grown to 150 acres, and was being cultivated by Rufus Cloud. Historical records from 1870 tell us that Cloud owned a horse, six cows, two oxen, seven cattle, and produced the following: 80 bushels of wheat, 100 bushels of corn, 60 bushels of barley, 50 tons of hay, 400 bushels of potatoes, 700 pounds of butter, 100 pounds of cheese, 450 pounds of wool, $25 in orchard products, $158 in forest products, and 600 pounds of maple syrup. Eventually the farm supported 100 head of milking cows, and supplied dairy products to Dartmouth College.

The early-middle part of the 19th century was the most prosperous agricultural era in Norwich and the state of Vermont. Merino sheep were introduced to Windsor County in 1819, and for four decades afterward, the land was cleared, stone walls erected, and sheep barns built. Eventually, wool and lamb production moved west, but with the advent of rail service, hill farmers could haul milk to the rail station for shipping to cities. Although the farm population declined, the soils on the Vermont side of the river grew good hay and continued to be farmed by those willing to work hard.

In 1921, a young couple from Somerville, Massachusetts, Bob and Betty McKenzie, along with Betty's father, Frank Wentworth, bought the farmstead. Bob McKenzie milked cows, while Betty kept house and worked in the bookbindery at Baker Library at Dartmouth College. Betty reminisced:

"Of course they were looking for good soil on which to farm, but how great it happened to be in such a beautiful spot. We had gardens, we had animals, we had chickens. We had everything that we ate, our vegetables, potatoes, apples, all meats, cream and milk, and eggs to cook with, so that we didn't have any trouble eating. So the farm was really self-sufficient as far as food went. There was always plenty of food. A lot of hard work, yes. In those days of course you had to can everything because you didn't have any other way of keeping it. No deep freeze in those days, and no refrigerator either, only icebox. But I canned it, and there wasn't any waste."

She continues: "Some time after we moved in, a neighbor gave us a copy of a film made by the artist Paul Sample. In it, Sample demonstrates to his Dartmouth students how he composes a painting. His subject was Bob McKenzie, and the setting for the portrait was the muddy barnyard with his herd. It gave us a glimpse of 1950s life on a Vermont hillside farm."

Evidence of previous gardening efforts was to be found all around the property. There were flowering shrubs and roses on all sides of the house, including a massive lilac (*Syringa vulgaris* 'President Lincoln') and a sour

LEFT Nearly a century old, a specimen of *Syringa vulgaris* 'President Lincoln' grows on the north side of the barn. **RIGHT** The rear of the house as it appeared in 1991, with collapsing deck and tangle of shrubs, roses, and rampant blackberries.

cherry (*Prunus* 'Montmorency') at the rear. Flowering plants also grew in beds in the lawn, though beds set against stone walls and the remaining barn were now empty. A remnant orchard lay across the road, with nearly a dozen apple trees for cider spread throughout. A large vegetable garden had gone to weeds in the back, but dozens of Mason jars of dill beans in the basement showed it had once been as productive as Betty said.

TAKING STOCK

After moving in, we began to take stock of what we had taken on. Old-fashioned shrubs, all of which had gone years without proper pruning, were planted on the east and south sides of the house. A thicket of roses infested with blackberries concealed the foundations; elsewhere, those prickly shrubs riddled a bed of *Rosa* 'Harison's Yellow' (also known as the "Yellow Rose of Texas"). A second massive lilac overshadowed the driveway, matching the one behind the house.

The yard on the east side was raised from the surrounding fields. Cut into the lawn here were flowerbeds of roses, quince, peonies, and one filled with

ASSESSING SITE CONDITIONS

WEATHER AND CLIMATE

The unpredictability and variability of winter temperatures and snow cover impact the garden more than absolute low temperatures. Bragg Hill is located in USDA Hardiness Zone 4b (-25 to -20°F), and we've recorded temperatures of -30°F on rare occasions in the nearly 30 years we've lived here, but it's an unusual winter when the mercury doesn't plunge below -20°F for at least a few nights in a row. Nighttime lows typically range from -10 to 35°F, and can vary widely within the space of one week. Daytime winter temperatures sometimes stay below freezing for days on end, but often top out around 35°F. Winter winds sweep down the valley from northwest to southeast.

Our average annual high is 56°F, with July and August highs in the low 80s, and increasingly reaching the 90s. Summer nighttime temperatures average in the mid- to high 50s.

Rainfall averages 44 inches per year, snowfall 82 inches, and precipitation runs slightly less in winter and spring. Snow cover, once reliable, now varies quite a bit. We often see snow on the ground before Christmas, but lose it in January during heavy rains. Sometimes the ground freezes at Thanksgiving, but with no appreciable snow cover until March. Snowy winters see accumulations of 3 to 4 feet on the ground, drifting higher near walls and buildings. Snow cover can linger well into April, but open ground typically thaws by the end of that month, and we are generally frost-free from May until late September.

Since the farm sits on a north slope, there are places where the ground freezes and remains frozen regardless of snowfall, especially at the north bases of stone walls. These areas also collect and hold snow, and often don't thaw before early May. More than anything else, winter plant losses result from voles and field mice tunneling under the snow to feed on the stems and roots of woody and herbaceous plants.

SOIL AND GEOLOGY

I am fortunate to garden on a nearly rock-free, mineral-rich, well-draining soil that is deep to bedrock. It's this combination of drainage and nearly neutral pH that makes the Vermont hills so green, and allows lilacs, fruit trees, hayfields, and many garden plants to thrive. The farm's soil is a loamy till from metamorphic rocks left by receding glaciers, the hillside gently sloping toward a brook that feeds the Connecticut River.

The contours of the present day Connecticut River Valley were shaped by Lake Hitchcock, a glacial lake formed during the last ice age when the river became obstructed at Rocky Hill, Connecticut. At the lake's edge, 160 miles north, sat Bragg Hill. Sediment deposited by the glaciers is responsible for the good soils found here today. Bedrock on our side of the river grew from deposits of crystallized mud, so the soil's pH is more neutral than the acidic soils on the New Hampshire side, which formed from granite.

AIR AND WATER CIRCULATION

Understanding where circulating air is warm or cool has aided in siting plants. Air patterns here are noticeable, so that on still, early autumn evenings, we can feel streams of cold air draining downslope. A particularly chilly flow drains along the west side of the Vegetable Garden, and it took me a while to pinpoint that as the source of reduced yields on that side. Many of the ornamental plants I grow, however, thrive in cooler nighttime temperatures. Our hillside also hosts wet seams where the water table emerges at the surface—a neighbor who has cut hay here for 35 years recalls accidentally plowing up hollow cedar logs that had been laid decades earlier to drain the field.

common tawny daylily and weedy raspberries—but most beds had been emptied of perennials and mulched over to await the next gardener.

The layout had the air of a classic northern New England farmyard, with its mix of shade trees, flowering shrubs, peonies, and flowerbeds. And although some of the shrubs were weedy and expendable, others were a bit unusual for the area, and told of a gardener having made smart choices. There was nothing out of the ordinary or ostentatious—this was a collection of the sturdiest of old-fashioned plants. One thing I surmised from this inventory of woody plants was that it was a pretty tough climate at the top of the hill, and that the protection of the house was necessary for any tree or shrub that wasn't entirely cold hardy. Historically, winters on Bragg Hill occasionally saw temperatures dip into the -25°F to -30°F range.

So the garden's layout made sense. As I took stock, I could see it would be worth cleaning out the brambles and pruning the shrubs while I thought about how to approach the landscape. I spent much of that fall pulling blackberries out from among the roses, and trying to determine what was worth saving and what had to go. Loppers in hand, I cut the long canes and piled them to burn, and then dug the roots out of the mass. Hardly any *Rosa* 'Harison's Yellow' remained—another couple of years of neglect and the roses would have been completely overtaken. Liberating them from the blackberries gave them room to spend the next few years filling in instead and growing once again to be a showpiece of the mid-June front yard.

I followed the blackberries around to the back of the house. After pulling out dozens, I discovered that part of the foundation had caved in, and the brambles had obscured the fact from the house inspector that there was no foundation underneath the kitchen and bathroom. That blackberry-pulling exercise led to the immediate expenditure of $10,000 for foundation work, but at least the bathroom didn't freeze that first winter, as it had surely done before.

On the west side of the house sat the one remaining barn. It had hand-sawn timbers dating back to the 1780s that would have been cut by the first owners. The roof had been kept up, and new foundations poured in the 1970s on what is a damp site. The other barns had collapsed or been carved up for parts, and by then were mostly piles of tin roofing, rotting posts, and mounds of decaying hay bales, littered with bits and pieces of barnyard implements.

The weeds in the rest of the yard pointed to compacted soil and neglect. Canada thistle, nettle, dock, and burdock grew out of carpets of ground ivy, plantain, and hairy bittercress. The vegetable garden had been overrun by quack grass.

The only one remaining from a complex of three, our working barn has timbers dating from the 1780s.

The site's native soil was heavier than any I had ever worked with. Although the vegetable garden had been improved over the years, each time I put my garden fork in the ground anywhere else, the cold, compacted soil was unyielding. I came to see the quality of this native soil—and its tilth when improved with organic matter—as one of the site's defining factors. Although plants may take an extra year or two to penetrate the heavy soil and gain a toe-hold, once they do, they grow robustly.

getting started

I spent most of the first fall getting to know the place. I would stick my garden fork in the ground, probing below thick tufts of weedy grass, to try to figure out what was underneath. More often than not, the fork hit tin roofing. Like the sites of the old barns, the yard was scattered with bits of tools and implements, with accumulations of old building parts, mounded with tufts of matted grass and burdock.

The old vegetable garden was a 45- by 64-foot rectangle, with earth pushed up on the lower side where the ground sloped away. I forked out some of the quack grass and sank a few plants I had carried with me from my former garden in nearby Union Village, mostly old-fashioned perennials I had salvaged from Cornish gardens. They were the makings of a first garden—a few historic bearded irises, old strains of garden phlox, and a couple of delphinium seedlings. The next spring, there appeared a most vigorous patch of rhubarb.

I spent hours forking over and picking out quack grass. The fork constantly got tangled up in chicken wire and barbed wire that Betty McKenzie had used to keep deer and woodchucks out of the garden. Metal posts tipped in every direction, with wire strung among them.

At that point in time, my unsolicited advice to a new homeowner who wanted to make a garden would've been to live with and observe your site for a year before taking any irreversible actions. This advice is easy to dispense but rather more difficult to practice. The idea is to study your property during the course of the seasons, to determine patterns of light and shade, seasonal wet spots, how the wind funnels in winter, and where the snow drifts. This gives you time to study why certain plants or existing features are the way they are, and allows you to gain a sense of what the place is like in each season—its

FAR LEFT The remains of a vegetable garden, with houses sprouting on parts of the farm sold off to pay taxes.

MIDDLE *Malus* 'Selkirk', a hardy crabapple with rosy pink flowers and bronze-tinged spring foliage.

LEFT Beauty bush and *Rosa* 'Harison's Yellow' make a striking show together in mid-June.

attributes and its challenges. This advice is also intended to give you time to think about how to use the land, to experience the moods it induces, and to find where you feel most centered in a property, where there is privacy, and where you feel exposed. Where do you want to take the sun, and where will you eventually want shade? This conservative approach permits you to take the time to figure out what others who came before intended, how and why things are the way they are, as well as what you can repair or enhance, or what ought to be removed or replaced.

I tried to envision what the place had looked like as a farm, and how I could make a garden out of its current state. The problem was not only the scope of the cleanup, but also the scale of the landscape beyond the immediate yard. How could I make a garden that made sense out of what was left of the farm-stead, as well as something compatible with the scale of the landscape that encompassed it? I endeavored to make a country garden—one that respected the heritage of the farm, yet would certainly be more elaborate than what any previous gardener would have contemplated.

I decided to carry on as much as possible with what Betty McKenzie had done in the front of the house. I didn't significantly alter the appearance of the garden as seen from the road. The front yard is actually one space that

LEFT *Rosa* 'Harison's Yellow' RIGHT *Paeonia* 'Festiva Maxima'

includes a large side yard formed by two long runs of stone walls. These walls have large, flat, weathered granite capstones, and are a defining feature of this part of the property. I had one section rebuilt that had collapsed. Long ago, perhaps when the house was first constructed, this yard was built up to facilitate drainage and provide a flat space for outdoor activities. Its lawn is the first to green up in spring.

A crabapple (*Malus* 'Selkirk') and apple (*M.* 'Yellow Transparent') complement our two mature sugar maples. What I like most about these small trees are the shadows they cast on the lawn at sunrise on hazy summer mornings. 'Selkirk' has red-tinged foliage and striking red flower buds that open to pink. Beauty bush (*Kolkwitzia amabalis*), *Spiraea ×vanhouttei*, Tatarian honeysuckle (*Lonicera tatarica*), and *Rosa* 'Harison's Yellow' clothe the east side of the house. The beauty bush and 'Harison's Yellow' rose bloom at the same time, and put on one of the greatest shows of the year, every year. Cut into a bed on the lawn is a row of white peonies, 'Festiva Maxima' and the intensely fragrant 'Duchesse de Nemours'. Between the 'Selkirk' crabapple and stone wall is a flower border that had daylilies and a short, vividly pink rose that I haven't been able to identify.

These were all plants that had survived years on a cold, windy hillside. Some of them had been shared with neighbors and showed up at other old farms on the hill. It's unlikely that I would have chosen any one of them if I were to have made the garden from scratch, but they've continued to give me pleasure and provide a certain reassuring sense of continuity for neighbors who walk by.

On the bank on the opposite side of the road (land that had once been part of the farm), there were remnant plantings of hydrangeas, roses, daylilies, and poet's narcissus, along with an orchard of old-fashioned keeping apples. Variegated goutweed had run through all of this.

I mostly took my own advice and didn't do much that was new in the first couple of years. I cleaned up the property by pruning shrubs, aggressively removing invasive and unwanted plants, and thinking through some of the practicalities involved in making a garden. I was busy with work elsewhere and didn't have a great deal of time to invest in starting a garden of my own. One thing I did decide to do was to screen the large house to the west. I was offered a few nursery-grown white pines at a bargain price because they had been damaged, and I planted them along a mound at what I thought would be the western edge of the garden. That planting soon failed. I learned two lessons: I should come to a better understanding of what the garden's scope would be before doing any further planting; and I should both invest in good plants, and only invest as much in plants as I could afford at one time.

envisioning the future

It became apparent that I needed to craft a master plan. I needed to ask myself what I wanted out of the garden, and how I wanted to organize it. This would involve mapping where the boundaries of the garden should be fixed, and where I would plant trees and hedges to establish a framework for the future.

I took measurements so I could draw the house, stone walls, and barn foundations on a base plan. I photographed the site and pasted shots on boards that I positioned in my work area, and I pinned a few inspirational snapshots of Wave Hill and other gardens to a corkboard. As I began to sketch ideas on paper, the plan fell into place. I had thought I would lay out a number of options, but as I drew the existing conditions, the lines of the garden emerged, and its scope became more extensive than I had anticipated. I knew that within the garden overall, I wanted a vegetable garden, some fruit trees, a small border in front, a larger flower garden in back, and plantings within the ruins of the barn foundations. I knew that I had to find a way to create a clear demarcation between garden and field, but also integrate garden, field, and landscape.

GUIDING PRINCIPLES

I also established guiding principles for the garden. Some were clearly articulated from the beginning, especially those having to do with practical matters, while others became important as the garden evolved and I became more conversant in its emotional and ecological aspects. Here are those principles, as I understand them today:

- The garden should feel like a natural outgrowth of the place, harmonizing with its vernacular architecture and agricultural character. It should relate to the topography and natural features around it, and establish a site line with Cardigan Mountain to the east.

- The front yard will remain largely as Betty McKenzie kept it, including trees, shrubs, and spatial organization. The empty flowerbed can be replanted as I please.

- The focus of development will be in the back, and should evolve out of the remnants of the former barns and existing vegetable garden. I should especially take advantage of the barn foundations for the distinctive spaces they could be used to create, and special plants that could be grown there.

- The best views are to the east and the northwest, and these should be featured and preserved. The long views of the hills to the north should be preserved if it's feasible.

There is need for an evergreen planting to screen the neighboring house to the west. A small apple orchard on a grid with existing apple trees should be planted in front of these evergreens. Deciduous screening is needed to the east in the eventuality of a new house, but planted so that views aren't obscured.

The garden should present well from the deck, and this view should attract the eye and invite exploration, but not give everything away.

There should be a sense of privacy in the back, and a quiet place to sit for respite from the road and the busier, more colorful parts of the garden.

Plants should be the primary means of creating structure and interest, rather than architecture or garden ornament. The garden should accommodate a wide range of hardy woody plants and herbaceous perennials.

There should be definite boundaries, and the garden should not expand into the field. The woodland edge along the north side of the field could be enhanced with judicious pruning and planting.

The garden should aim for outstanding quality of design, plant selection, and maintenance, but should not give the impression of striving for perfection. It should be a country garden, with straight lines where they matter, sensuous curves, and elements of formality, but relaxed and exuberant in feeling. It should evolve and not stand still. It should be informed, enchanting, and never complete. It should be interesting in all seasons.

The garden should be designed and maintained according to ecological principles and sustainability as I understand them, and enhance habitats for the benefit of wildlife.

Native plants should be employed when they are the best choice, and in abundance, to build an environment rich in native flora and fauna. Permanent plantings should be as deer-resistant as possible.

Plants should be grown as well as possible without recourse to chemical fertilizers or pesticides. The vegetable garden should be entirely organic, with inputs limited to compost and organic fertilizers.

Garden maintenance should be of a caliber to take pride in. The garden should be able to be maintained by myself with one other person working one day a week, with assists from tree and lawn maintenance professionals. It should be a healthy working environment for all.

The garden should offer a variety of experiences, and the long, narrow site should allow for variety in circulation for utility and pleasure. There should be arresting visual moments within the garden, not just long-range views.

The garden should make an emotional impact—on me and on visitors. The main purpose of the garden is for my own health and well-being, and the pleasure of working in it through being outside and absorbed in the natural world.

The garden should reflect my enthusiasms, interests, and personality.

I will share the garden with others.

I decided to treat the Vegetable and Flower Gardens as one unit, with an axis running through them, and poplars to separate them from the field.

DECISIONS TO MAKE

Working on paper identified three main problems. The first was that the overall potential garden space was long and narrow, and that the foundations of the former barns did not line up with the house. Second was that this narrowness made for an awkward transition to the landscape beyond. Third, it was going to be difficult to locate a vegetable garden anywhere other than right below the deck, in full view of the larger garden and landscape beyond.

I resolved this last problem by deciding that the new vegetable garden, contrary to prevailing wisdom, would be placed in that prominent position just below the deck. This move was appropriate for a number of reasons, but especially because it was a reasonable distance from the kitchen and water source, and for continuity. At the same time, I slated the large, rectangular space that had been the McKenzies' vegetable garden to become my main flower garden. A new area for vegetables by the deck could be scaled so it felt like a continuation of this space.

It made sense to give the Vegetable Garden pride of place because I wanted to retain the farmstead character of the property and reflect my background as a market gardener. I admired well-tended vegetable gardens and felt that I could make this one interesting and productive.

This decision allowed me to create an axis that ran through both the Flower and Vegetable Gardens. Both have a central path that lines up with the peak of Cardigan Mountain (although the mountain became less a focal point when new neighbors built a house to the east). I planted an apple tree on axis 150 feet to the west, anticipating it would one day provide another focal point. I positioned an arbor at the western entrance of the Flower Garden, which frames a view to the now-mature apple tree. This conceit worked well enough for the first few years as the garden matured—there was an obvious connection with Cardigan, though that connection has been obscured as trees and shrub borders have grown. Even so, the internal order that this established has remained.

The Flower Garden would eventually be filled with a dense mix of perennials and shrubs. For the moment, a long grass path with an arbor at one end and a stone entry step at the other, along with a shorter cross-axis grass path, established an orderly framework and helped draw an axial connection that runs through the Vegetable Garden, Flower Garden, and long lawn framed by apple trees. This arrangement was similar to many images of old-fashioned flower gardens I was familiar with from books and photographs, and is not unlike the flower garden at Wave Hill.

Screening the neighbor's house was necessary. We enjoyed our neighbors, and I think the feeling was mutual, but we could discern their TV preferences from the garden. I looked around the region to see which evergreens were most effective as screens over the long term, and it was easier to see what didn't work. Native white and red pines lost their lower limbs as the trees grew and upper limbs shaded out those below. Native balsam fir and arborvitae were hardy, but prone to deer browse, especially fir, which would have been my first choice—had it been an option, a fir planting would have tied into an existing group of trees on the neighbor's property.

Although Norway spruce have a dark, somber quality, I chose them because they seemed to offer the greatest long-term potential. They often hold their

CLOCKWISE FROM ABOVE One tree was planted on axis with the path through the garden. Mown paths in the tall grass encourage a closer look. • Norway spruce makes an effective screen, but its somber presence called for a small apple orchard underplanted with daffodils. • 'Liberty', 'Cortland', and 'McIntosh' apples planted on a grid, with gaps, to make the new garden feel part of the farm.

lower branches into old age, as I had noted at a number of other sites. Norway spruce are entirely hardy, and not susceptible to deer browse. As of this writing, they have worked well for 25 years, although they have begun to thin at the bottom due to a newly prevalent needle cast disease.

A landscape architect friend suggested the addition of a small orchard of apple trees would be in keeping with the farm landscape, as well as provide additional screening, making the utilitarian nature of the spruces less apparent. He recommended I plant them on a grid based on two existing apple trees. A landscape contractor I knew had access to a number of 15-year-old 'McIntosh' apples that were being culled from an orchard. He dug and transplanted a half dozen of these trees, and I added 'Cortland', 'Macoun', and 'Liberty' saplings to round out the orchard. Now, when the apples bloom in May, they pop even more with the dark evergreen spruces behind them. The orchard has become a focal point in spring with the addition of hundreds of yellow 'King Alfred' type daffodils planted in the tall grass between the trees. We mow paths among the apples, as well as one long sweeping path following the spruces to the neighbors' backyard, and shorter loops returning to

different parts of the garden. I've encouraged milkweed for monarch butter-flies among the tall meadow grasses.

My original plan envisioned a row of shrubs between the far side of the Flower Garden and field, with two columnar arborvitaes acting as a gateway into the field between the Flower and Vegetable Gardens. As soon as the arbor-vitaes were planted, it was apparent what a foolish decision that was. No mat-ter their eventual size, the formality of such an evergreen was simply wrong.

The solution came when I visited Frank and Anne Cabot's Les Quatre Vents in Quebec. Frank founded The Garden Conservancy in 1989, with encourage-ment from Anne, and repeated visits to their garden have had a profound effect on mine. At first, I didn't understand or appreciate the way Les Quatre Vents unfolds in a series of experiences, some almost spiritual, others humorous. But I kept going back, and roaming unescorted, and in time came to appreciate the emotional journey this garden offers. On my first visit, I noticed how Lombardy poplars, nearly a century old, formed its backbone. These trees connected the house to long hedged gardens and views beyond. I also saw how well they worked as a distinctive landscape feature in the Saint Lawrence River Valley. With Frank's encouragement, I decided to plant Lombardy poplars along the rear edge of my Flower Garden. This separated the garden from the field, created a backdrop, and would lead the eye to the verdant hills and blue sky beyond.

I chose *Populus nigra* var. *betulifolia* 'Theves' (syn. *Populus nigra* var. *thevestina*) because I had experience planting this cultivar at Aspet. Its habit is a bit wider than the classic Lombardy poplar, but it's less prone to disease and longer-lived than the typical Lombardy. I planted a row of seven, spacing them two each behind each section of the Flower Garden and Vegetable Garden.

The poplars have given us years of enjoyment, and although I might have chosen a more sturdy, long-lived tree, I get a lot of satisfaction from the sound and sight of their leaves in motion in even the slightest breeze. They are a great presence from the deck and porch where, when seated, you see nothing of the Flower Garden below—only the sentinel rhythm of the poplars with framed views between them of the horizon and sky beyond. Their impact is the oppo-site from the field below, where they read as sentries, lined up in a row.

Poplars are not without their problems, but what has made them a success in my garden was Frank's advice to cut them back severely when they reached 20 years of age. My poplars grew quickly and tall, and were effective for a number of years before eventually beginning to thin out. At that point, they had reached almost 60 feet in height. I later topped them at 16 feet, and now they've flushed out again, and are on their way to becoming handsome speci-mens like before.

Poplars frame views and create rhythm, bringing life to the garden.

SPIRIT OF PLACE

I'm told that my garden became more Italian in feeling after I spent some time at the Villa Serbelloni at Bellagio on Lake Como. Certainly the poplars contributed to this notion as they matured, but what I also observed at the Villa Serbelloni was how successfully that garden transitioned: first, a villa with terraces and outbuildings; next, formal gardens with evergreens, shrubs, and perennials; then fields and orchards with olives and fruit trees; and finally the majestic landscape of lake and mountains. I spent hours roaming that garden and this made an impression—perhaps unconscious at the time—that reinforced my desire to make a garden to stroll in, and for there to be places to pause to take in views, as well as a variety of experiences with no one prescribed route.

This may be part of the reason the garden expanded in scope beyond my original intention. The farther one gets from the house, the more interesting the views, and the more the sense of being in nature prevails. Likewise, the

This retaining wall was rebuilt and clothed in small-leaved climbing hydrangeas, with paper birches above and a foliage border along its base. OPPOSITE As plantings matured and additions were made, the garden became more reminiscent of Italy or France.

potential for me to make interesting garden scenes intensifies. I don't want to give the impression that I was after a villa-style garden; I also looked closely at small-scale country gardens in Italy, and was especially drawn to those that combined a bit of formal structure, through the use of walls or hedges, with productive and ornamental plantings. I brought home indelible memories of gardens composed of a mix of hedges, fruit trees, roses, and perennials, along with rows of herbs and vegetables.

To the west of the barn was a 75-foot-long retaining wall. Part of this wall had collapsed, and I saw it could be rebuilt to shore up a terrace above, while doubling as a dramatic backdrop for a border below. I decided to enclose the flat area below the wall with a hedge, creating a large open space immediately adjacent to the Flower Garden and affording views of the fields and orchard.

On the terrace above the wall, in what had been the yard for the milk cows between the barns, I planted a grove of paper birch, not unlike a grove of white

A grove of paper birch, recalling one at Aspet, took root in the former barnyard.

FAR RIGHT Under the enveloping branches of birches and poplars, a view to the hills and fields to the north.

birches I'd observed at Aspet. A bench is set toward the edge of the lawn for taking in the view. This is the place where we set up food and drinks when we entertain groups of people outside. I enjoy watching guests linger in the pleasant shade of the birches.

At the edge of the yard was a single-story milking parlor that had collapsed decades earlier. Beyond lay stone foundation walls that supported a stable. The wood frames and siding of all three structures had been carted off for salvage; the remnants of the stable had been repurposed in the woods farther up the road as a cabin for a group of back-to-the-land types in the 1970s. What was left, including lots of tin roofing, was scattered about under the shade

of successional apple trees, aspens, and a 50-foot-tall elm. The milking parlor's concrete slab suggested a place for shade-loving plants and ferns, and eventually a rock garden, while the stable foundation was perfect for a *hortus inclusus*, or enclosed garden. Even though it was possible to envision these as garden spaces, they were all in need of regrading, clearing out debris, and repairs to their walls.

The purpose and plantings of the ruined milking parlor have evolved as trees and plants I've tried there have failed or succeeded. It began as a shady nook for ferns, but as trees died and were removed, it has become the sunny, colorful Rock Garden, where I experiment with plants and collect some of those that do best. The Stable has become the place I seek out when I want to be alone with my thoughts. Its floor is covered with junipers and willows, bringing a bit of the landscape's distant mountaintops into the garden.

Beyond the Stable, paths lead to apple trees and the field, and to newer parts of the garden, some of which I'm still thinking through. As the trees I planted in the early years begin to reach maturity, new possibilities arise, and new vistas open up. Parts of the garden have changed, either intentionally, because of storm damage, or as a result of occasional losses, and the inevitable changes have had to be dealt with. At the same time, I have been able to correct some initial mistakes. In the intervening years, horticulture itself has changed, and we must now squarely face challenges with climate change and the disruption of so many natural cycles.

My initial impulse to create a variety of garden experiences came primarily from the desire to create beautiful spaces with plants, but I couldn't have anticipated at the beginning how my horticultural interests would develop over the years. I began the garden with limited experience, and while the work I did and lessons I learned in other gardens contributed to my garden's design, it also contributed to the plant collections that inhabit it today.

Adam Gopnik, writing in the *New Yorker* about Alexander Calder, says, "Influence among artists works in many ways, but usually the most potent is to provide not a series of patterns but a set of permissions—confidence that what one is already inclined to do is not trivial." For me, the permission to trust my collector's impulse took hold with the old-fashioned perennials I grew working in the Cornish gardens. Historic bearded iris, garden phlox, and New England and New York aster were some of the first plant types I collected. Many came from those gardens, but in time, I began to assemble collections from nursery sources and other collectors. The collecting mentality eventually extended to a range of plants: willows, saxifrages, primulas, hellebores, rhubarbs, rodgersias, hydrangeas, epimediums, ferns, gentians, sedges, and ornamental grasses.

I experiment with plants, and I grow them so I can make knowledgeable choices when designing and planting other people's gardens. I work to incorporate a range of trees, shrubs, perennials, and grasses that grow well in other regions but may be untested locally. My goals vary—in some cases I'm interested in knowing what attributes some underused or underrated plant might

The cracked foundations of the former milking parlor became a setting for rock garden plants.

The Stable is a retreat for privacy, its floor carpeted with junipers and willows. OPPOSITE Horticultural interests and inspiration evolve in unpredictable ways.

offer. I've been especially curious to figure out what native shrubs perform well in garden compositions, and I have collected a variety of native perennials to assess and select the best performers. I'm also interested in understanding winter hardiness, rates of growth, ultimate size, and habit, especially for shrubs. I'm thankful to gardeners in other regions who have pressed upon me plants they think should succeed, or at least be tried, in my northern garden.

One of my guiding principles for the garden is that it should be maintained by myself and one other person who works one day each week during the growing season. Susan Howard has been that person from almost the beginning, and I wouldn't have made the garden without her. Sue is a mainstay, and it is a far better garden because of her great eye, collaborative personality, and extraordinary work ethic. She is my sounding board for new ideas, and I listen to her when things aren't as they should be. In practice, her patience and meticulousness with grooming, staking, and deadheading elevates the garden from a dense assortment of perennials into the realm of garden art.

Over the course of 25 years, the garden has expanded and been reconfigured as I've collected, and as Sue and I have recognized possibilities. Looking back, I'm amazed at how quickly it came together—what it means and where it's headed are often on my mind. Its design, however, has remained paramount. My first goal is and has always been to create a beautiful, memorable setting, and then populate it with interesting and appropriate plants. It is to create a sense of place that imparts an emotion. Whether it's serenity, drama, or excitement, the emotional response comes first, and the plants must emphasize that response. As the garden has matured, a narrative quality has emerged, and I'll explore that in the chapters that follow, as well as look at more of the influences in its making, and how my intentions and circumstances have changed over the years as the garden has evolved.

a new old-fashioned flower garden

The Flower Garden is at the heart of the garden overall, and from the beginning, I envisioned it as the stage where the main show would take place.

Columns of *Thuja occidentalis* 'DeGroot's Spire' and 'Emerald Green' lead the eye to the landscape beyond.

The Flower and Vegetable Gardens as seen from the deck. **OPPOSITE** The garden rests comfortably in the landscape and leads the eye toward Moose Mountain to the east.

Its generous size and its shape were set by the footprint of what had been the previous owners' vegetable garden. It is an ample space, but one that answers the challenge of creating a garden appropriate to the surrounding landscape. Any effort has to be substantial to be proportionate—and the enormity of the view suggested a massing of shrubs and perennials.

The site for the Flower Garden is 65 feet long by 45 feet wide, and best appreciated from the deck. In planning its layout, I divided the rectangle into four quadrants, with a long grass path as a central axis running east to west, and a shorter cross axis running north to south. I wasn't interested in building walls or fences, and I wanted to make it a garden where plants formed the structure. Even though I call it the Flower Garden, it is a mixed planting, with hardy shrubs, roses, vines, perennials, bulbs, and a few annuals. Although I did consider hedging it, I decided I wanted a garden that emerged out of the landscape and did not feel imposed upon it.

The house and garden sit on an open slope that gradually descends toward the woods and neighboring valleys and hillsides. The land on which

the garden sits slants away from the house, so I knew it would be important to make the garden feel grounded, and not as though it were sliding down the slope. Our hayfield comes to within 100 feet of the back of the house, and I decided early on that I wouldn't extend the garden into that field. It's beautiful throughout the seasons and full of life, with nesting bobolinks and woodcock in spring, as well as the occasional coyote or moose—and (it goes without saying) deer.

For inspiration, I had a head full of images of Arts and Crafts and Colonial Revival–style gardens, some of which I found in Louise Shelton's *Beautiful Gardens in America* and Guy Lowell's *American Gardens*. Closer to home, Stephen Parrish's Northcote was particularly vivid for me, having researched the garden and done a partial restoration. Ellen Biddle Shipman's full and beautifully composed flower gardens also left an impression, especially her borders at Aspet, as well as others I'd seen in photographs. Other flower gardens I knew firsthand, like Wave Hill and Hollister House, had the benefit of architecture—the former with its conservatory, and the latter with its tall

CLOCKWISE FROM LEFT Ellen Shipman's flower gardens informed my use of evergreens for accent and old-fashioned summer-blooming perennials for color. • Stephen Parrish's Northcote embodied a number of ideals I aspired to. • As a landscape painter, Charles Platt approached his garden and the landscape as a unified composition.

brick walls and massive yew hedges. I found these inspiring, but couldn't contemplate their architectural ambitiousness for my garden.

At nearby Northcote, Stephen Parrish used trees and shrubs—some clipped into hedges and others allowed to reach their full potential—to create a garden space between the house and a long view over the valley. Parrish made a garden to live in, to cultivate, and to take pleasure in as he moved through his daily life.

In addition to its scale, and its role as a framework for plantings, I wanted this part of the garden to connect to the landscape so it felt a part of the countryside and open to the sky. Following Northcote's example, and images of Shipman gardens, I decided I would add upright conifers, small trees, and cutback shrubs into the composition to facilitate that. Another Cornish designer hinted at a way to preserve the view. In laying out High Court, where the sight of the river valley and Mount Ascutney dominated, Charles Platt insisted that the view remain open and unencumbered by a flower garden, so he placed it to the side of the house, and treated the foreground to the view very simply. I had little choice but to site my Flower Garden in front of the view, and for that, I found precedent in Platt's own garden in Cornish. He laid out three terraces of geometric beds directly between the house and the view down the river valley, outlined with paths leading toward the vista. Platt's experience as a landscape painter prepared him to tie the garden to the landscape through plantings of undulating shrubs and vertical accents of conifers and Lombardy poplars.

My Flower Garden is laid out in a simple rectangle, and is bisected by two paths, the longer of which lines up with Cardigan Mountain to the east, and an apple tree planted in tall field grass to the west. A single granite threshold serves as an entrance from the Vegetable Garden. Divided into four quadrants, this area was originally edged on the east and south sides by herbaceous peonies; on the west side with a row of tall shrub roses; and on the north with a mix of shrub roses and other deciduous shrubs.

Key to the Flower Garden's planning was my desire to achieve a sense of unity, harmony, and repetition in how it's viewed from the deck and other vantage points. The interior of each quadrant is planted with shrubs for substance and height. Taller perennials are repeated throughout the four beds to tie it all together and encourage the eye to keep moving through the interplay of habit, shape, texture, and color.

I give the eye something to fix on—often an upright form, like a columnar arborvitae—but couple that with activity to draw the eye away. Large, rounded shrubs and drifts of perennials with foliage and flowers in complementary colors accomplish this. I've repeated form and color in strategically placed shrubs, some with deep burgundy foliage, for example, as well as the silvery foliage of willows. Taller, long-blooming perennials weave throughout, some with seed heads that look just as good as the flowers, each contributing in some way, bringing the entire composition into harmony.

From the deck, the landscape is a vignette of rolling hills and fields with garden foreground. Following Platt's example in his garden, the upright forms of Lombardy poplars and the dark spires of arborvitaes—similar to the Italian cypress in Tuscan gardens—tie this part of the garden to the landscape. Practitioners have created similar compositions at such sites as Villa Serbelloni in Italy, Les Quatre Vents in Canada, and Great Dixter in England.

The Lombardy poplars are spaced to allow views through, creating a lofty green backdrop for the Flower Garden. Their foliage moves in even the slightest breeze, captures the eye, drawing in the viewer from distant parts of the

garden and the field. They are equally effective whether from a distance or from within the Flower Garden. Up close, their massive, fissured trunks lend the garden a sense of structure and steadfastness.

The columnar spires of various arborvitae cultivars echo those of the poplars, acting as intermediaries in scale, and like them, lead the eye to the landscape beyond: *Thuja occidentalis* 'DeGroot's Spire', 'Emerald Green', 'Holmstrup', and 'Malonyana'. Their dark-needled foliage complements that of Norway spruce in the middle distance and stands of white pine on distant hillsides, as well as the burgundy foliage of surrounding shrubs. The arborvitaes emphasize the verticality of the poplars and the spruces, and make a striking connection between the two. Their footprint isn't sizeable, nor do they cast much shadow, and this allows me to grow a variety of sun-loving perennials beneath and around them.

The poplars are equally effective up close or from a distance.

FAR LEFT
Dark-needled arborvitae help tie the garden to the landscape beyond.

a fresh start

In July 1996, Sue and I set aside a day to attack the Flower Garden's initial framework. We lifted and moved around dozens of perennials that came either as divisions or had been growing in rows as nursery stock. We began with repetition, using the same shrubs and perennials in each of the four quadrants. The project began to come together.

All of this work was done by hand. I shoveled the occasional load of composted cow manure onto the garden. The granite slab, which serves as a threshold, was muscled into place, and the peonies and roses planted. I had placed the two paths, the border of peonies and roses was taking hold, and other shrubs within the beds were putting on some growth. But the garden wasn't yet working as a composition, and had no background to set it off from the field beyond.

Shrubs with dark foliage like *Berberis thunbergii* var. *purpurea* 'Royal Cloak' and purple-leaf sand cherry are balanced with silver foliage of rosemary willow and *Salix alba* 'Silver Column'.

The arborvitaes and poplars were a start. To this, I added shrubs with dark foliage: *Berberis thunbergii* var. *purpurea* 'Royal Cloak', purple-leaf sand cherry (*Prunus ×cistena*), and later *Sambucus nigra* 'Black Beauty' and *Gleditsia triacanthos* var. *inermis* 'Ruby Lace'. One early choice, *Physocarpus opulifolius* 'Diabolo', was later removed because it underperformed compared to *B.* 'Royal Cloak'.

These shrubs are striking when viewed from above. Within the garden, a number of lower-growing burgundy-leaved barberries create their own repeating pattern, as well as echoing the taller barberries, like 'Royal Cloak'. The best of these lower, more mounded cultivars is 'Concorde'. I grow them all for the color and quality of their foliage and not for floral interest, the one exception being the brilliant orange flower of 'Royal Cloak', which is stunning in contrast to its burgundy leaf. Neither 'Royal Cloak' or 'Concorde' has shown any tendency toward invasiveness here.

Willows have played an important role in lightening the somber tones of that dark foliage. Most effective are rosemary willow (*Salix elaeagnos* subsp. *angustifolia*), treated as a cutback shrub, *S. alba* 'Silver Column', and the hybrid Russian olive *Elaeagnus* 'Quicksilver'. As landscape designer Harland Hand writes in the *American Rock Garden Society Bulletin* about his California garden, "The trick is to place these strong dark or light shapes rhythmically through the middle-tone areas. The more the dark and light contrast dominates . . . the more brilliant and jewel-like the remaining colors become and the better all colors seem to go together."

Light, silvery perennials on the ground plane include *Stachys byzantina* 'Helene von Stein', *Artemisia* 'Valerie Finnis', an unidentified artemisia related to *A.* 'Silver Queen', and the light green of *Sedum* 'Autumn Joy'. In flowering, this sedum also provides the benefit of picking up on and repeating the barberries' mounded shape, a trick that is reinforced as the season progresses and temperatures cool. The flower heads of 'Autumn Joy' deepen to a maroon that faces down the darkening tone of barberry.

A silver-leaved vine that draws a lot of attention is *Lonicera prolifera*, which I purchased from Cady's Falls Nursery—sadly now closed—but it is very similar to a plant on the market called *L. reticulata* 'Kintzley's Ghost'.

Other shrubs planted throughout the garden that help tie it together are panicle hydrangeas, mountain hydrangeas, and shrub roses. I planted a number of panicle cultivars before the two that show best here became available on the market: *Hydrangea paniculata* 'Quick Fire', my favorite, and 'Pinky Winky'. Both bloom early and quickly fade to pink. I like the height they give, and that they can be kept in scale through annual pruning.

I have only recently introduced mountain hydrangea (*Hydrangea serrata*) into the garden. Where I can give it room, this plant is a most welcome addition. I was given a small piece of the cultivar 'Blue Billow', and planted it on the central walkway's edge. It has increased in size and beauty each year, yet remains fairly compact compared to some other hydrangeas. The foliage is smaller than *H. macrophylla*, a sort of felt green in summer, tinged with red in fall. This hydrangea's lacecap flowers are deeper blue in the center and lighter outside, fading to pink, and it blooms on new and old wood. It's proven to be entirely hardy, so the original planting of 'Blue Billow' has led me to experiment with other closely related mountain hydrangeas in the border. *Hydrangea serrata* 'Annie's Blue' (descended from 'Blue Billow') and 'Blue Bird' also get good marks. Their solid, mounded shape and manageable size is welcome along the edges. They can help prop some of the looser, more freely growing perennials too, and I sometimes use them as support for clematis.

A couple of shrub roses have proven themselves worthy for flowering, habit, and substance. *Rosa* 'William Baffin', one of the Canadian Explorer series, grows in my garden both as a climber and as a shrub. I needed a tall shrub rose that I could also train over the arbor at the west end of the garden. 'William Baffin', in my experience, is the most robust hardy rose that suits both purposes. It has trained well on the arbor, and I've let it bulk up into a 10 foot by 10 foot shrub that provides a backdrop for perennials in its quadrant, and screens the open lawn beyond the arbor.

I've also planted *Rosa* 'William Baffin' together with purple-leaf sand cherry and encouraged the bright red rose to grow up through the cherry's purple leaves, rendering an eye-catching combination. This planting fills a lot of problematic real estate in the northeastern quad, on a slope that

FAR LEFT Silvery *Lonicera prolifera* twines through an arbor at the west end of the garden.

MIDDLE Panicle hydrangeas at different stages in their progression from white to pink in the late August border, accompanied by *Elaeagnus* 'Quicksilver'.

LEFT Reliably blue lacecap flowers and burgundy fall foliage make 'Blue Billow' mountain hydrangea a multiseason mainstay.

dries out quickly, where the rose commingles with the cherry and panicle hydrangeas, forming a bulwark at that end of the garden that requires little maintenance. It's pleasing to see the bright pink blooms of 'William Baffin' at both ends of the garden, whereas I think a single large group could draw too much attention. The repetition of plants and color throughout the quadrants through the course of the season remains one of the hallmarks of the Flower Garden.

A much gentler presence than *Rosa* 'William Baffin' is the shrub rose *R.* 'Frau Dagmar Hastrup', a rugosa with delicate, single, silvery pink blossoms. Though it's reputed to have a continuous season of bloom, mine produces one early flush, and occasional sparser blooms in late summer. Its flower is followed by a showy hip that resembles a large crabapple.

Another rugosa I grow is *Rosa* 'Pink Grootendorst', with its small, fringed pink clusters that grow taller and have a more pronounced repeat bloom. I have discovered that many roses don't last long in a densely packed garden, but 'Pink Grootendorst' has persisted, and although it requires regular pruning, it continues to send out tall flowering spikes year in and year out.

Two other roses have proven long-lived here. One is *R. alba* 'Königin von Danemark', which has large, beautifully formed, deliciously fragrant blooms in soft pink, coupled with gray-green leaves on thorny stems 6 feet high. The second, *R. glauca*, promises bluish foliage with purple highlights, soft pink flowers, and orange hips that fade to chocolate.

A number of perennials help knit the garden together through their persistent foliage. Some of these also have a fairly long season of flowering. The strong, robust growers whose foliage remains good throughout the garden season include Joe Pye weed, Siberian iris, monkshood, pink turtlehead, New York aster, sedum, lady's mantle, hardy geranium, and ornamental grasses. There are a few perennials with an exceptionally long blooming season, and these are repeated throughout the four quadrants: Martha Washington's

FAR LEFT *Rosa* 'William Baffin'

MIDDLE *Rosa alba* 'Königin von Danemark'

LEFT Robust perennials knit the garden together, including phlox, Joe Pye weed, and *Sedum* 'Autumn Joy'.

plume, meadow rue, and cultivars of bee balm, veronica, garden phlox, and allium.

Phlox is one of the biggest contributors to the Flower Garden. It conveys substance when seen from above and within, and its long flowering season comes in a variety of complementary colors. The flower of *Eupatorium*, or Joe Pye weed, echoes the mounded shape of the phlox, only with a coarser leaf and taller stature. It is slow to get going in spring, and can flag during midsummer's hot days and dry spells, and its foliage may be chomped by caterpillars or slugs. Even so, it grows out to produce large, late-season inflorescences alive with pollinators, especially monarch butterflies. Joe Pye weed's flowers fade to a dusky seed head that in the best varieties complements its strong purple stems, as well as fading flower and foliage of neighboring panicle hydrangeas.

The flowers of two other tall perennials are significant in the way they weave the garden together. The tall, spikey flower heads of rhizomatous *Filipendula rubra* 'Venusta Magnifica'—commonly called Martha Washington's plume, or queen of the prairie—run throughout. Early in the

Other tall perennials, like delphiniums and *Filipendula rubra* 'Venusta Magnifica', tie the four quadrants together.

OPPOSITE Lady's mantle is one of a handful of low-growing perennials used at intervals along bed edges.

season, their soft pink tufts complement the blues of delphiniums and veronicas. They're too pink for my taste at times, but as these flower heads age, they deepen to a dusky tone that is the perfect complement for the purples and mauves of late summer. The wind and their neighboring plants play with them, and they lend an informal, natural air to the more steadfast shrubs and perennials. Meadow rue (*Thalictrum rochebruneanum* 'Lavender Mist') performs a similar function. Its 9-foot-tall misty flower stalks are like a spray in a flower arrangement—they protrude above the solid green foliage of their neighbors, sway in the breeze, and bring the blue of the sky into the garden.

Three particular monkshoods of height have earned a place in each quadrant: the early *Aconitum napellus*, midsummer *A.* 'Spark's Variety', and October-blooming *A. carmichaelli*. Another repeated perennial is a mid-size old-fashioned *Veronica longifolia* cultivar brought from Aspet, which has relaxed spires of gentle blue, and complements the pinks and blues of clematis and Martha Washington's plume. *Campanula lactiflora* 'Prichard's Variety' is a recent acquisition I'm now increasing and using throughout. Cultivars of *Nepeta* have a place in each quadrant too, along with hollyhocks, when they are

Geranium dalmaticum with *Artemisia* 'Valerie Finnis'.

FAR RIGHT
Nepeta, sea holly, and rose campion, backed by bee balm, *Filipendula*, and *Hydrangea arborescens* 'White Dome'.

given room to self-sow or accept transplanting. *Veronicastrum* and *Vernonia* both hold their own; neither is particularly spectacular in this setting, but they are both often covered with pollinators.

I have planted delphiniums in each quadrant, and from late June through July, these plants carry the Flower Garden—this is especially true when it's viewed from a distance. They are at home in the loamy, neutral soil here, and often reach 7 to 8 feet in height, and can last for nearly a month.

The garden's perimeter, especially the southern edge seen from the approach, is bordered with three low-growing perennials: lady's mantle (*Alchemilla mollis*), *Geranium dalmaticum*, and *Sedum* 'Autumn Joy'. Rising between these lower growers are the strapping leaves of Siberian irises, as well as dense, taller foliage of garden phlox and New York aster. A couple of *Hydrangea serrata* lend substance to this front-of-the-border composition.

The western edge of the garden is the only portion that I have made somewhat formal by planting a low yew hedge that frames the vine-covered arch. The eastern edge of the garden slopes to the lawn and is bordered with lamb's ears (*Stachys byzantina* 'Helene von Stein'), ground phlox (*Phlox subulata*), and *Dianthus* 'Bath's Pink', with nepeta, sea holly, and rose campion rising just in from the outer edge. Large shrubs were the theme of the north side of the garden for a number of years—*Salix* 'Matsudana', large *Elaeagnus* 'Quicksilver', and native elderberries. The willow recently failed, and the Russian olive has

been cut back to encourage fresh sprouts. I've taken this opportunity to make that side more interesting by adding hydrangeas, *Clethra*, and *Persicaria*.

Within the garden, I continue the use of low-growing plants along the interior paths, punctuated by taller, upright foliage. The long axis is defined by *Berberis* 'Concorde', *Iris* 'Caesar's Brother', *Sedum* 'Autumn Joy', *Geranium dalmaticum*, *Heuchera* or coral bells, bearded iris, and allium. The cross axis changes things up a bit, and I've put in slightly taller plants, like silvery gray *Artemisia* 'Valerie Finnis' and 'Silver Queen', *Hydrangea serrata* 'Blue Billow', *Geranium magnificum*, and *Deschampsia cespitosa*.

I enjoy observing weekly changes along the length of the central path. Lamb's ears hug the granite step, with their gray foliage picked up by artemisias and sedums. The upright spikes of the bearded irises and the rounded forms of the red-leaved barberries add contrast in both color and shape. Upright arborvitaes complete this composition. The bed's interior is filled with dense blocks and drifts of perennials, grown closely together to make the most of their flowering, as well as to shade out weeds. All of the perennials are entirely hardy and gain in size every year; this requires ongoing assessment to maintain a balance, but the idea is that once the design has been arrived at, it can be fairly stable. That doesn't mean that I don't occasionally remove a plant and reshape a bed, but the goal is to fill the garden with plants that perform well and combine well with one another.

progression through the seasons

I've never felt constrained by the fact that I garden in Vermont and am limited by a short growing season. I seek out something of interest and beauty in the garden throughout the year, and enjoy its changing moods and a progression of featured plants. I'm especially interested in orchestrating compositions full of color and visual appeal from early May into late October. I do this for my own satisfaction as much as for anything else; the only way to accomplish this is by being in the garden, working it, observing, and making small adjustments as I go along.

SPRING

The arborvitaes, shrubs, yew hedge, and metal arch, backed by the poplars, give the Flower Garden its structure in winter and early spring. But every day, as the snow melts and the ground begins to thaw, I see new growth in perennials pushing out of the ground and the stems of willows and other shrubs coloring up as sap begins to push out buds. Monkshoods and delphiniums are poised to unfurl in the still-frozen ground in late April.

My first tasks of the garden year are to prune any winter-damaged or broken wood out of shrubs. The first pass is to remove damaged wood, then older wood or excessive growth is removed, and finally cuts are made to give the shrubs a pleasing shape. Those that receive an annual cutting back, such as rosemary willow, are pruned and shaped at this point.

Some years this work feels burdensome, especially if any of the thorny shrubs has gotten out of scale or died back excessively. But for the most part, I take pleasure in shaping the shrubs, and enjoying the complementary forms of freshly pruned hydrangeas, roses, and willows bordered by the fresh green of the grass paths. For a garden as loosely arranged as this, the geometry of the grass paths and the yew hedge provide something of a frame, and the upright shapes of the shrubs and conifers help ground the composition. Lower, mounding shrubs create a rhythm along the paths, and are a foil for the surrounding loose growth of the grasses and perennials.

Pruning and shaping roses, barberries, and willows
is the first task of the garden year.

Spring brings other chores, nearly all of them welcome after a winter of disengagement from the garden. There's always some cleanup to do that wasn't completed in fall. Asters and bee balms do not winter well if cut to the ground in fall, so that task waits until spring. Clematis needs pruning, last year's flowers are clipped off the hydrangeas, and the roses shaped. If the boundaries of beds weren't edged in fall, then spring is the time to do that, but it's an easier task in fall when the ground is drier and soil warmer. Any weeds that have found their way in are removed. There aren't many annual weeds in the garden, but there are a couple of problematic perennials that have gained a toehold, and this is when we go after them. Goutweed (*Aegopodium podagraria*) was a problem, but it has been eradicated; lily-of-the-valley (*Convallaria majalis*) and star of Bethlehem (*Ornithogalum umbellatum*) are more pressing concerns today. All of this has to be worked around changeable weather and soil conditions—some parts of the garden dry out earlier than others and can be worked on, while other parts remain soggy, and any foot traffic will result in compacted soil. Cold days and early mornings when there is a crust of frost still on the ground are a good time for pruning and cleaning up, while any real digging must wait until the soil has begun to dry out. It usually takes a couple of weeks of intermittent work to tidy things up; a fine raking finishes spring cleanup.

Certain robust perennials like phlox and Joe Pye weed need to be contained, and one of the ways I do this is to take a spade and dig out a third or more of the plant before they are in active growth, usually taking out a portion of the perimeter so I can get at some of the older portions at the core, where the plants tend to die out. I usually just flip the sod over and fill in with a bit of compost. This is also the time to look for phlox that has seeded around. These are forked out, the soil shaken off the roots, and the plants pitched onto the compost pile. Anything large enough to dig and worth giving away is handled that way.

The high point of spring tasks is the fine-tuning of the planting scheme. The previous fall, I will have gone through and made notes about improvements I have in mind, combinations to try, failed experiments to undo, and more thoroughgoing renovations to plan for. Some of this work can be done in fall, especially removing those plants that have outlived their usefulness. Choice plants that could find a place in another part of the garden, or someone else's garden, wait for spring to be dug and moved. But in fall I will have decided what changes I want to make, and I keep a running tally of notes of plants to move, and partnerships I want to encourage. At times, I will draw a quick sketch to indicate how the plants will be positioned. I type this up in a

The blue of camassia is just different enough from allium to make it a pretty picture.

document on my computer that I update as tasks are completed. To keep track of details, I may also write instructions on a wooden or plastic label and slip it into the ground next to the plant in question, sometimes also marking the spot with a bamboo stake. The combination of notes on paper, a label in the ground, and a stake marking the spot usually helps me get the job done without too much hesitation in spring. This helps me identify the plants I want to move, their locations in my nursery area or in other parts of the garden, and to transplant them as soon as the ground is workable.

All of this means I can move quickly in spring, shifting plants around in the garden before they have put on much growth. It works extremely well for shrubs that want to be moved before they leaf out, and for perennials that have barely broken ground. A big benefit of moving this early in spring is that it requires less watering to establish plants, and although they may grow a little smaller than normal the first season, they will have sufficient root systems to compete with neighboring plants. I do not like planting container-grown plants directly into the Flower Garden, but rather purchase them (either by mail or from nurseries) and grow them in a holding bed in the Vegetable Garden for a year or two. I move them once they've had a chance to get adjusted to the soil and bulk up their root system. My goal is to have any transplanting done by the end of April.

By the middle of May, the garden has pulled out of mud season and taken on the vitality of spring: there is a rhythm and harmony that comes from the emerging dark burgundy foliage of the barberries and elderberries, and the silvery gray-greens of the willows and hydrangeas. The dark greens of the peonies and Siberian irises, lime greens of sedums and delphiniums, and the grassy green of the paths make for a vibrancy in the beds that is unmatched at any other time of year. The composition of the garden reads clearly and cheerfully.

There is no reason to dilute this effect with the addition of smaller, early spring bulbs such as squill or narcissus, and so I have held off introducing spring bulbs into the Flower Garden. I use them in other beds, but there is clarity of color and composition that comes just from the foliage of the emerging shrubs and perennials that is fixed by the deep permanent greens of the conifers. The eye can move from plant to plant, appreciating each for its own habit, leaf color, and shape. And the foliage is entirely healthy, not yet having been scorched by the sun or eaten by insects. The lush tops of the delphiniums, the tips of the peonies, and the barberries' emerging reds may be just as fleeting as their later flowers, but it is, for me, a more satisfying moment in the garden.

EARLY SUMMER

The first display of flowers is a wash of blues and purples from *Camassia leichtlinii* 'Caerulea' and *Allium* 'Purple Sensation'. The round heads of the alliums float just above the tops of expanding perennial foliage. I began by planting a hundred allium bulbs spaced equally among the quadrants; they are self-sowing and have scattered themselves throughout the beds, and as a result, taken on the appearance of having occurred naturally. Camassia is a newer addition, and the cerulean blue of its spire-shaped flowers opening from bottom to top is just different enough from the globe shape of the alliums to make it interesting.

Some of the offspring of 'Purple Sensation' are a lighter mauve, and that contrast helps them blend into the green of the plants' foliage, while the deep purple of the majority is a foil for the silvery tones of the willows and mounds of burgundy 'Concorde' barberries. This combination continues over two weeks in late May and into the first half of June. It is a prelude to the roses and

The planting plan shines in early June.

June-blooming perennials that are gathering momentum, and shrubs transitioning from their fresh juvenile growth into mature growth.

It's at this season that the Flower Garden's planting plan seems most intentional. Each plant shows its individual, characteristic growth habit. Low, rounded shapes are repeated along the paths and the narrow, strapping foliage of irises and alliums offers contrast. The darker green foliage of the Siberian irises sets off the lighter greens of sedums, geraniums, lamb's ears, and artemisias. The shrubs can be appreciated for their graceful habits and rhythm. By mid-May, shrubs and perennials have leafed out sufficiently to form a structured composition, while the alliums and camassias float among them. This modulation continues into June, when the bulbs have faded, and the purple wings of the Siberian iris cultivar 'Caesar's Brother' hover above its spikey foliage. This is also when the foliage of Siberian iris is at its most useful—dozens of tall, dark green spears hold to each other and break up the mounds. There was a period when I had given up on Siberian iris, because its leaves tend to overgrow their space and collapse late in the season. But the garden was lacking the note of upright, narrow foliage, and the Siberians are the best of the hardy irises for this purpose. Other varieties of Siberian iris have not been reliable bloomers for me, so I've divided and repeated 'Caesar's Brother' in key locations. Its handsome bloom also continues the purple theme begun by the alliums.

Harland Hand cited the importance of massing tones of green in the garden, including light tones, middle tones, and dark tones. I can see I was pursuing some of the same ideas in my choice of plants for foliage. The light tones found in artemisia, lamb's ear, rose campion, bearded iris, lady's mantle, *Dianthus*, and *Sedum* 'Autumn Joy' are useful to place along a path. Rosemary willow, 'Silver Column' willow, and *Elaeagnus* 'Quicksilver' broadcast their taller silvery foliage from a distance. Middle tones are found in the leaves of hydrangea, phlox, aster, and Joe Pye weed, while dark tones are pronounced in conifers like arborvitae and yew, and the garden's purple deciduous shrubs—elderberry, sand cherry, honeylocust, and barberry.

For light-colored flowers, I favor the blues of delphinium, bearded iris, clematis, veronica, and *Browallia*; the light pinks or lavenders of *Filipendula* and meadow rue; and the chartreuse of lady's mantle. I try to avoid white because it's too strident compared to the rest of the planting. Although a number of panicle hydrangeas are allowed that start off white, they fade to dusky pink. Middle colors include the mauves of Joe Pye weed and allium, while dark colors are provided by bee balm, monkshood, rose, hollyhock, New York aster, and the flower of *Sedum* 'Autumn Joy'. Various cultivars of garden phlox offer a range of light, middle, and dark.

After the subdued harmony of the garden in late spring, there comes the burgeoning colorfulness and lushness of its mid-June iteration. *Rosa spinossissima* is the first of the roses to bloom, and its fragrant, single white blossoms are an immediate attraction. Then the lavender mops of bearded irises and the delicate pinks of heirloom coral bells initiate a season of an old-fashioned (and decidedly passé) combination of pinks and lavenders that is the motif of June. Only as the heat of summer intensifies does the garden move beyond this "grandmother's garden" effect. For the first part of the season, many of the perennials and their combinations are straight out of Helena Rutherfurd Ely's 1902 book *A Woman's Hardy Garden*, one of the first to be published in this country on perennials for garden interest. I've learned of alternatives for creating mixed borders that require less meticulous maintenance and offer an extended season of interest from the gardens of other gifted practitioners. I've also experimented with more contemporary ideas and styles in my work in

Paeonia 'Sarah Bernhardt'

other gardens. I believe there is still a certain honest beauty in this once fashionable style of gardening.

These old-time standbys were the plants I first became familiar with working in Cornish gardens, and I continue to appreciate them in my own. Other ideas and many newer plants have taken a place here, and in gardens I make for other people, but many of the plants and practices that gardeners like Ely or Ellen Shipman recommended continue to be sound from the standpoint of design, maintenance, and even ecology. The challenge of working in this type of garden is often the question of labor; in the past, this style was popular when owners could typically hire a part-time gardener to help. Although Sue and I do put in time in the Flower Garden, we keep it to a reasonable level, and are happy to do so because I believe this style of mixed border is an appropriate type of garden for the setting, and can evolve as new ideas, plants, and techniques come along.

The knowledge and skills I've built working in this garden allow me to help others create gardens based on hardy perennials and shrubs. I look for plants that give a long season of interest, and for planting combinations and techniques that can be effective in a variety of situations. But this garden is also a reminder that gardening really does require a certain amount of skill and physical work, and that's just the way it is—as with any discipline. Good gardening requires a familiarity with plants, along with a passion to create something to bring joy into our lives.

The Flower Garden's first iteration was bordered on three sides with peonies and on the farthest side with tall shrub roses. That was never entirely

successful—not all the shrub roses were hardy or manageable in this setting. Some peonies took hold, but I mistakenly allowed others to be overtaken by more vigorous perennials. *Paeonia* 'Sarah Bernhardt' is the most elegant among them; *P.* 'Duchesse de Nemours' is a lush and fragrant one.

The garden reaches a certain peak in late June as the days lengthen and evenings linger. The last of the peonies and roses are in the full flush of pink and white, and will soon be coupled with the tall spires of delphiniums. The lavender of nepetas, pink of coral bells, and fresh silvery gray of artemisias, lamb's ears, and sedums reinforce the theme. The Flower Garden is aglow when backlit by the lowering sun.

The edges of the beds are a haze in late June, with small flowering perennials and biennials blooming in pink and purple: *Geranium dalmaticum*, *G. magnificum*, *G. renardii*, and *G. sanguineum*; in addition to *Allium karataviense*, *Campanula* 'Joan Elliot', *Salvia verticillata* 'Purple Rain', *Clematis integrifolia*, artemisia, and the still-vibrant foliage of the barberries.

Along the area in the front of the garden, a rather compact form of lady's mantle with foamy chartreuse flowers alternates with *Geranium dalmaticum*, sedums, and Siberian irises. The geranium sports a bright pink flower that sputters along, but I use it here mainly for contrast in height—it is one of the

FAR LEFT Flowers of *Iris* 'Caesar's Brother' hover above spikey foliage, and play off mounds of *Sedum* 'Autumn Joy' and *Berberis* 'Concorde'.

MIDDLE Allium, lady's mantle, artemisia, *Geranium magnificum*, and *G. dalmaticum* along the cross axis.

LEFT Descendants of my original planting of three hybrid Karl Foerster delphiniums, accompanied by lavender *Campanula lactiflora* 'Prichard's Variety' and *Rosa* 'Pink Grootendorst'.

hardiest lower-growing compact perennials that is reliable on the heavy soil here, and it is nearly evergreen. Sedums and irises break up the mat of geranium and lady's mantle.

Early on, I sought out a number of *Nepeta* species and cultivars. Their lavender haze spreads through the garden from June onward. They're not especially adapted to being so crowded, but some have found a permanent place, especially *Nepeta sibirica* and its cultivar 'Souvenir d'Andre Chaudron', which work their way up through other perennials, and, with deadheading, last much of summer. *Nepeta kubanica* is another good one.

Late June sees the ascent of delphinium. I first worked with delphiniums in the borders at Aspet. They did well in that garden's fertile soil, where they received adequate moisture and enjoyed cool night temperatures. The surrounding hedges protected them from squalls, and the high canopy and dappled sunlight of neighboring white birches added protection. There were two robust clumps of the Belladonna Group that grew in the garden for decades. I took small divisions of these and tried new strains as well. The Pacific Hybrids series did not perform exceptionally well, with the exception of 'Black Knight'. I raised seedlings from other seed strains and would move them into the garden once they had attained some size. Our white pine hedges were a beautiful

LEFT Blue spires of Karl Foerster hybrid delphiniums make good company with pink phloxes. RIGHT *Campanula lactiflora* 'Prichard's Variety'

background for the sky blue spires of delphiniums—the garden was worthy of a painting during their season. One admirer remarked that she hadn't seen such lavish use of delphiniums since visiting the Connecticut garden of artist Edward Steichen.

My friend Joan Platt shared a few seeds of a Karl Foerster hybrid delphinium gathered from a garden in the Berkshires. Three germinated, and I set them out in the nursery. At that point, I wasn't familiar with the hybridizing work of Foerster, a great German nurseryman. Each of the three plants grew into distinctive hybrids. One was tall and sported a sky blue flower with a perfect center, also called a bee. Another was shorter, but had glossy, dark green foliage and a purple flower. The third had a lighter leaf with a light blue belladonna-like flower, also with a perfect bee. These three were exceptionally long-lived for delphiniums, the longest thriving for 17 years. The first to succumb, ironically, was the one with the stout foliage—it perished during a summer of extended heat and drought, after blooming and being cut back. I collected seeds from all three and have grown dozens of successor plants. Occasionally they would seed themselves into the garden, and I have followed

Joan's example and passed along seeds to nurseries and other gardeners. I have tried, without success, to obtain new seeds of Foerster Hybrid delphinium in order to reinvigorate the strain, but I did acquire another first-rate German hybrid, 'Volkerfrieden', known for its electric blue flowers. This strain is also long-lived.

If we're lucky, cool evenings and moist, foggy mornings alternate with the hot and sunny days of late June. It is the cool evening air that makes the Flower Garden thrive. Without this cooling after the heat of the day, the perennials here lose their robustness and their saturated flower color. Vermont gardens have few advantages over those in more temperate parts of the country, but perennials such as delphiniums that need the respite of cool night air after daytime heat will thrive here if given the chance. It remains to be seen whether these plants will continue to be stalwarts of Vermont gardens, as gardeners in other parts of New England are giving up on plants that require cool nights. For now, most seasons here are suitable for cultivating these finicky plants, but in some years now I do lose some delphiniums to heat and humidity.

It used to seem that as the delphiniums were fading and their stems snapped during a squall, the most refined moment of the Flower Garden had passed. But there are other midsummer perennials that, though perhaps not as attention-grabbing, make their own statement and combine well in smart compositions.

The blue clouds of *Campanula lactiflora* 'Prichard's Variety', a relative newcomer to the garden, are foremost among them. This campanula caught my attention a few Julys ago in a garden on Mount Desert Island, Maine, and it's a classic cottage garden plant. 'Prichard's Variety' grew 5 feet tall and 4 feet across in that maritime climate. Seeing it in Maine prompted me to seek it out, and I found it at Digging Dog Nursery, one of the few nurseries offering it. Though it's located in Mendocino County, California, Digging Dog has a deep list of hard-to-find perennials that happen to do exceptionally well in my climate too. *Campanula* 'Prichard's Variety' begins to bloom with the delphiniums and enriches the July garden with its pale blue. Pinching back a few stems in each clump when they're in bud promotes a longer season of bloom.

Mid-July is a peak time for the old-fashioned northern New England flower garden. Cornish Colony gardeners would strive for an abundance of bloom at this time—delphinium, campanula, Carolina lupine, nepeta, rose, phlox, hollyhock, astilbe, geranium, hydrangea, clematis. The trick is to keep the garden going after that main show.

Moving deeper into July, the vivid pinks and roses of hollyhocks stand tall well into August, and can be just as distinctive a garden element as delphiniums. Hollyhocks can be a challenge to grow well, as their foliage is prone to rust in some years, and they can be knocked back by a snowless winter, but when they flourish, they carry the garden to loftier heights.

Aconitum 'Spark's Variety' also picks up where the delphiniums left off, and meadow rue creates a cloud of lavender and yellow that floats above the garden for weeks on end. *Filipendula* opens cotton-candy pink, but luckily for my taste, fades to a more tolerable straw color.

The Flower Garden has improved as I have brought new ideas and interesting plants back from my travels. I first noticed *Berberis* 'Royal Cloak' at Bellevue Botanical Garden in Washington state, and I was impressed by its velvety burgundy leaves and orange blossoms. I tucked three plants in 4-inch pots into my backpack to carry home on the plane, and today, this cultivar is part of the garden's backbone. Another such plant is *Sambucus nigra* 'Black Beauty', an elderberry cultivar. Having first seen it for sale in the United Kingdom, I then noticed its consistently rich dark foliage in other gardens, and was finally able to locate one to purchase on the West Coast. Along with that plant, I brought back an idea: I would grow purple-flowered *Clematis* 'Polish Spirit' through the dark purple foliage of the elderberry. This combination reaches another level entirely when 'Black Beauty' opens its clusters of pink flowers. I've planted *Allium* 'Millenium' at the elderberry's base for its pinkish globes, and sprinkled about the small true-blue flowers of *Browallia americana*.

Sue and I have learned that some of the taller monkshoods, delphiniums, and veronicas can use an assist to keep them upright, and we've fashioned hoops out of rolled rebar, or remesh, to surround the plants in June so they remain standing after the squalls of July and August. We do this rather than stake plants, because it not only saves labor, it's far more effective and more aesthetically pleasing than bamboo stakes and unsightly twine. These supports go in as the foliage of individual plants starts to grow together. The supports are barely visible, as the rebar is rusted and its dark color blends into green foliage; it is eventually entirely obscured. Peonies get a similar treatment with shorter hoops also made from rolled rebar. The wire is a narrow gauge, and we cut it so each section can be bent into a hoop, with prongs to press into the soil. Sometimes we drive a stake through the mesh to stabilize the hoop in the ground. We wouldn't be able to grow some of these taller plants without some method of staking, and it takes far less time to set up and take down compared to bamboo stakes and twine.

When the garden was first planted, I experimented with annuals, especially *Nicotiana*, *Salvia*, *Amaranthus*, and *Verbena bonariensis*. All combine well with perennials. In recent years, I've curbed their use in favor of the most

LEFT Bee balm, *Filipendula*, and a strain of *Veronica longifolia* collected from a Cornish garden, with *Clematis* 'Perle d'Azur' running through them. MIDDLE An old-fashioned hollyhock strain passed down by local gardeners. RIGHT *Clematis* 'Polish Spirit'

reliable and longest blooming. My favorite is little-known *Browallia americana*, a tropical forget-me-not native to South America. In May, I dibble in two or three dozen transplants along the front of each quadrant, weaving them between low-growing perennials for support as both gain in height. *Browallia americana* self-sows from the previous year's stock, and by August a fresh crop appears to carry through until hard frost. The soft blue flowers of this species are far more delicate than the variety of *Browallia* that is widely available as a bedding plant.

I also pop in one or two dozen *Nicotiana langsdorfii*. Its small, tubular greenish yellow flowers complement any medium or darker-colored flowers and foliage. It too reseeds, and volunteers begin flowering the second half of the season, but I tend to get more mileage out of starts. Some of the self-sowing annual poppy cultivars will also cluster where there is open room along the garden's margins, especially a stunning purple-flowered opium poppy. Rose campion (*Lychnis coronaria*) colonizes along the front of the border as well. I rely on it for its silvery, felty foliage that lights up the edges of the paths for the

entire season, and the brilliant magenta of its flowers that plug away for weeks. I've given up on *Verbena bonariensis*, as the garden is too crowded for it to show itself to its best advantage.

Summer-blooming phlox flowers in the garden from late June to October, and it's indispensible in the border in late summer. If properly sited, with adequate moisture, dozens of heirloom and contemporary varieties do well in the New England garden. Garden phlox is a reliable performer, entirely hardy, long-lasting in bloom, and quite fragrant—the nighttime garden is potent with its scent. I've collected and grown dozens of varieties gathered from gardens and nurseries, observed them, and selected the best to feature here. Those prone to mildew are rejected—I will not spray to control mildew. Sometimes I give a plant a second chance, but any that shows mildew in succeeding years gets pitched. Some cultivars suffer from spider mites, and in that case I try shifting them into cooler, moister locations where they'll be less stressed by heat and dryness. Lately, I'm focused on collecting darker, richer pinks, reds, and purples, as well as later-blooming varieties that extend the show well into fall.

When I first made the garden, I built a collection of summer-blooming phlox. Many came from Cornish gardens, which usually meant they were varieties that were 50 to 100 years old, some among the earliest named cultivars. Rachel Kane has assembled an extraordinary compendium of these phloxes at Perennial Pleasures Nursery in East Hardwick, Vermont. She is the go-to person for gardeners looking for heirloom varieties, and an expert at tracking them down, identifying them, and assigning them accurate names when possible.

I ended up with far more phlox than I wanted to grow in the Flower Garden alone, so we opened a 70-foot-long bed along a stone foundation wall in perfectly rich soil of rotted hay and cow manure. Here, I grouped my collection by colors, with red, orange, and stronger colors at the far end, lavender and purple in the middle, blending into pink and light rose near the end, and white sprinkled throughout.

The first to flower is a long-blooming, long-lived heirloom variety called 'Rosalinde'. Its magenta-pink panicles appear in late June and combine well with lady's mantle, artemisia, and rose campion. 'Rosalinde' is derived from the species *Phlox maculata*. Those phloxes born of *P. paniculata* begin blooming toward mid-July, and hold the middle and center portions of the beds

Silver tones of *Elaeagnus* 'Quicksilver', *Aconitum napellus*, *Hydrangea* 'Pinky Winky', *Sedum* 'Autumn Joy', and phloxes in a late July composition.

RIGHT Soft violet-blue flowers of annual *Browallia americana* harmonize well with silvery foliage of artemisia and *Clematis integrifolia* 'Rosea'.

MIDDLE I've permitted phlox to occupy a chunk of real estate along the back of the border for its fragrance and cheerful, long-lasting bloom.

FAR RIGHT Early blooming *Phlox* 'Rosalinde' and the grass *Calamagrostis ×acutiflora* 'Karl Foerster' in bloom, backed by a seed-grown Foerster hybrid delphinium.

together. They are positioned both for the cheerful color they bring to the party, and because they hold their own in the density of the beds.

Although I gave up on trying to build a comprehensive collection, I have continued to acquire mildew-resistant varieties that bring a strong presence to the garden, because I prefer vibrant colors and medium-size, stout, clean foliage. A friend passed on an exceptionally long-blooming selection of heirloom phlox originally from Meadowburn, the garden of author Helena Rutherfurd Ely in New Jersey.

LATE SUMMER

It's usually in late July or early August that the garden gets its first thorough-going cleanup. The delphiniums have finished and their flower stalks are cut to the ground, except for those whose seeds I collect. Lady's mantle is dead-headed, cut back, and any discolored leaves removed. Campanulas are tipped back to remove fading flowers and promote newer buds below. Some excess

Siberian iris foliage is culled. Nepetas are deadheaded, cut to a node just above new buds, especially to promote rebloom; *Salvia verticillata* 'Purple Rain' is similarly pruned. Any remaining spent flower stalks of *Allium* 'Purple Sensation' or camassia are gently tugged and removed. Previously staked, taller perennials, such as monkshood and veronica, may have to be disentangled from neighbors and set back upright. This is also the time when we deadhead roses, and more vigorous shrubs like the rosemary willow are pruned to remove excess foliage and touch up their shape.

Drenching rains and gusty winds are the primary cause of plants toppling over. Many right themselves as they dry out in the sun. Those with larger flower heads may need an assist, which I do with a pair of long bamboo poles that I slip under the foliage and use to shake off moisture and tease the plants back into place.

The garden begins to show some gaps, with delphiniums or damaged plants cut back, and yellowing foliage from disease, or snail and slug damage. A close inspection reveals these shortcomings, and the garden can feel jungly from within with the towering monkshoods and Joe Pye weed, but this is when

it shows best from the deck. It seems too much to ask to keep a garden this densely planted immaculate. What is possible, however, is to keep it presentable from the deck all summer long. The farther the garden goes into the growing season, the more cutting back we do, and the better it looks. It may seem a difficult choice to remove the last blooms of nepetas or delphiniums, but once they're clipped back, the garden takes on a fresher, cleaner appearance for the remainder of summer and fall.

Panicle hydrangea is a hallmark of the late-season Flower Garden. I have paired these with a number of late-blooming perennials like veronica, meadow rue, and *Veronicastrum*, as well as with clematis planted at the base of the hydrangeas. The blues, violets, and purples of these vines weave well through some of the panicle hydrangeas, most of which open white but fade to pink. By late July, Joe Pye weed makes its presence known with the first flush of its ruddy pink flower heads.

In August, the garden becomes deeply saturated in color, with red and purple bee balm, deep pink phlox, purple monkshood, and the panicle hydrangeas.

In late August, phloxes, bee balms, Joe Pye weed, monkshoods, and meadow rue peak with the pre-autumnal seedheads of *Calamagrostis* and *Filipendula*.

FAR LEFT In midsummer, the garden gets a thorough cutting-back of spent foliage and flowers, and tall plants shored up to prevent them spilling onto neighbors.

Not only are the colors more saturated than the pinks and blues of June and July, there is a profusion of color, sometimes to the point that it's overwhelming. By mid-August, the garden reaches a crescendo of bloom. Nearly all the phloxes are open, and the dusty rose flowers of Joe Pye weed smother its foliage.

The Flower Garden also undergoes a mood shift in August. The solid blooms of phlox, monkshood, and hydrangea are softened by the flowers of ornamental grasses, notably *Calamagrostis ×acutiflora* 'Karl Foerster', whose awns introduce an earthy brown note tinged with dusky pink. This complements the darkening pink flower heads of *Filipendula* and the shorter tufted hair grass, *Deschampsia cespitosa*, whose arching flower stems catch the light of the afternoon sun.

This is also the time when the red and purple foliage of the barberries, elderberries, and *Gleditsia triacanthos* var. *inermis* 'Ruby Lace' really pulls its weight. These plants are a perfect foil for August and September perennials. The nights usually begin to cool in mid- or late August, and with that, the dark foliage of these shrubs becomes more saturated in color.

FALL

The garden used to start winding down in late August. With the first round of phloxes gone and the monkshoods and meadow rues over the hill, the garden lacked the excitement of strong color and shrewd plant combinations. But one of the lessons I learned at Aspet is that a presentable flower garden in northern New England is possible into October. It requires proper plant selection and routine grooming. In recent years, I have added more late summer color, especially late-blooming phlox, such as the deep red *Phlox* 'Nicky' and *P.* 'Red Magic', as well as the red spikes of *Persicaria* 'Blackfair'.

New York ironweed (*Vernonia noveboracensis*) enjoys a late but fairly brief flowering season, though its deep green foliage remains healthy and upright.

LEFT By late September, most summer perennials have passed, but Joe Pye weed, aster, and ironweed soldier on. RIGHT After the hard frosts of October, flowering ends, but foliage persists. OPPOSITE An heirloom variety of New York aster with *Calamagrostis* ×*acutiflora* 'Karl Foerster'.

I've added a couple of newer cultivars of Joe Pye weed, which are shorter, with darker flowers and stems. Pink turtlehead (*Chelone*), is being given more room, and I'm increasing the numbers of New York asters and fall-blooming monkshoods.

Early September brings another red flush of the striking *Rosa* 'William Baffin', and the softer pink of *Rosa* 'Pink Grootendorst'. *Sedum* 'Autumn Joy' starts to color up and the hydrangeas go dusky. Joe Pye weed takes on a mauve haze. By the middle of the month, the heads of the sedum are as red as brick, and the seed heads of 'Karl Foerster' feather reed grass have gone to straw. At this point, I thin some of the queen of the prairie's plumes so their wintry, brown appearance doesn't depress the rest of the garden. The color palette turns toward deep blue and purple late in September with the first flowers of New York asters.

One of the most successful and long-lasting combinations in the garden is that of the tawny grass *Calamagrostis ×acutiflora* 'Karl Foerster' and the purple daisies of New York aster (*Symphyotrichum novi-belgii*). Placed on the leading edge of the garden, where it can be enjoyed from a distance, this duo takes center stage for fall. To my mind, the combination says a lot about the value of preservation: the old (and for now nameless) heritage selection of New York aster pairs with *Calamagrostis*, named plant of the year by the Perennial Plant Association in 2001.

Other asters add a soft touch to late-season color, like *Symphyotrichum* 'Little Carlow' and *S. cordifolium* 'Chieftain', both upright and reliable blues. The blue daisies of *Aster ×frikartii* 'Monch' can bloom from July to frost.

It is this time of the year, with its medley of brilliant oranges and reds and waning buffs and browns, that we are reminded how compatible the garden and landscape are. The reds of sedum and *Filipendula*, and the brilliant scarlet of the adjoining cranberry viburnum hedge, sing out to the brilliant oranges, yellows, and reds of red maples and birches. The yellowish, fading foliage of Joe Pye weed, Siberian irises, and russet of grasses look to the sugar maples and poplars. Burgundy of the barberry and silver of the willow deepen considerably and harmonize more intensely than at any point earlier in the season. The upright conifers reassert their architectural role. Asters contribute their variety of blue and purple. The cycle of the seasons progresses with seed heads and rose hips, and a few remaining flowers visited less frequently, but appreciated nonetheless.

October's shorter days and longer, chillier nights finally bring a close to the growing season. In recent years, the first killing frosts have come in the middle of October, and finish off the flowers of all but the hardiest asters, phloxes, and monkshoods. Foliage begins to collapse, and Sue and I begin to cut back the Flower Garden in stages, removing the drab faded blooms of early monkshood, meadow rue, the worst of the diseased or fading flower stalks, and those that have snapped or blown over, like Siberian iris. I try to cut the garden back as late as possible and leave those plants standing that serve a purpose for wildlife. But it does need to be cut, otherwise it would be a soggy mess in spring. It's also a relief to see the architecture of the garden revealed again after the rampant lushness of summer growth. The roses and shrubs hold it all together, and the longer we go without cutting it all the way back, the more pleasure it brings, and the more sustenance it might provide for migrating birds and nectar-loving insects.

DECK PLANTINGS

The deck is home to a changing collection of container plants. When it was built, two recessed sections were included in the seat that skirts it. At one time, I arranged bright red geraniums in the recesses; now the longer section is for seating, and a shorter section at a corner is for a small collection of succulents. At the hotter end by the porch, larger pots of agaves, Jerusalem sages, and lavenders enjoy the baking sun. At the eastern end, which enjoys morning sun and afternoon shade, a changing collection of flowering and foliage plants rotates each summer. Those grown for foliage, like figs, artemisia, ornamental sedges, and *Bergenia* are overwintered in the cellar. Flowering plants and annuals, some chosen especially for fragrance, are purchased each year, including heliotrope, petunia, geranium, *Osteospermum*, fuchsia, begonia, and zinnia.

the front border

I sometimes wish the Flower Garden could be more charged with clashing colors, screaming with oranges and scarlets, invigorating the more saturated hues of pinks, purples, and blues of the perennials—the sort of garden Christopher Lloyd and Fergus Garrett of Great Dixter would feel at home in. For now, I will stay the course in the main garden, but tucked off at the far end of the front lawn is a border where I'm indulging those thoughts and playing with more color.

Originally planted with shrubs and old-fashioned perennials, this 60-foot-long space has a traditional feel to it, with an early spring massing of daffodils, a mid-June display of bearded irises and peonies, and an early summer appearance of roses, delphiniums, and *Thermopsis caroliniana*, or Carolina lupine. A north–south running stone wall separates the lawn and

MIDDLE Daffodils planted in groups of 10 between clumps of perennials and stems of shrub roses.

TOP LEFT *Thermopsis caroliniana* and delphinium both open their flowers from bottom to top, and look good from every angle.

BOTTOM LEFT *Delphinium* 'Volkerfrieden' is long-lived and reliable. Its foliage obscures a metal support.

border from the field; on the lawn just to the west of the border is a 'Selkirk' crabapple tree. Its deep pink blossoms draw attention in May, but unfortunately its spreading limbs shade the bed in the afternoon.

This border is also one of the easier areas to maintain—and sometimes feels like it may become the most satisfying. Part of what makes its care relatively easy is that it is anchored at either end by two large shrubs, *Viburnum sargentii* 'Onondaga' and *Hydrangea paniculata* 'Grandiflora'. A generous clump of rhubarb, one of the few perennials remaining from the previous owners, spreads along the wall under the hydrangea. It's a standout in bloom, but sadly its foliage collapses if I allow it to flower. Some years it gets to make a statement—others I'm grateful for the big, bold display of its rugged green leaves. Roses weave their way along the wall and through some of the perennials. Taller perennials are arrayed in large blocks against the wall, and mid-size ones positioned toward the middle and front.

It took a while to figure out that many of the perennials, such as daylilies, were heliotropic, meaning they turn their flowers toward the sun. The problem was they faced east, away from the lawn, the house, and the viewer. In dealing with this miscalculation, I've been trying to make a more coherent planting with perennials whose flowers are arrayed in panicles or clusters that do not turn to track the sun—for example, mulleins, veronicas, and mallows.

I am revving up this border with a more exciting range of perennial color than in the Flower Garden proper. High summer now sports a brighter show of pinks, with phlox and *Malva*, blue and purple with campanula and veronica, red and purple bee balm, tall yellows of cup plant and *Helianthus*, and hits of orange from daylily and *Helenium*. Later summer–blooming New York asters, feather reed grass, and monkshoods wind the season down. Each new addition of contrasting pinks, reds, and oranges makes my heart race just a little.

Large stands of cup plant, *Helianthus*, phlox, and *Helenium* can go all summer without being touched. (I do pinch back the taller growing *Helenium* cultivars in late spring to keep them in scale.) A mix of gooseneck loosestrife and *Monarda* hold the far end of the bed, with the hydrangea keeping them from working their way into the rest of the planting.

Low-growing clumps of sedum, geranium, veronica, peony, and a mountain hydrangea help hold the front, with upright bearded iris and daylily adding vertical spikes. A few gaps are held along the front of the border for some valuable self-sowers to move around from year to year, the most persistent being *Verbascum chaixii*, a double feverfew, and teasel.

Both this and the main Flower Garden feel compatible with the house and the setting, and have a sense of belonging. English-style perennial borders do not succeed in many parts of the country, but for now they are still doable in northern New England. This is a country garden set around an old farmhouse, and an old-fashioned garden packed with old-fashioned plants feels right. I've made gardens in other styles, and with other plants, but this is the heart of the greater garden; it took on this flavor more than two decades ago, and it is wearing well— with some modest updates.

CLOCKWISE FROM TOP LEFT Mullein, veronica, mallow, and delphinium flowers are arrayed in panicles that do not turn away from the viewer to face the sun. • *Viburnum sargentii* 'Onondaga' showing some of its maroon-tinged spring foliage and dramatic flower clusters. Red fall leaves and berries add a second season of interest. • Large, deeply ruffled leaves of *Rheum* 'Victoria' and its 5- to 6-foot-tall flower stalks in June. • A heat-loving group for August: *Helenium*, cup plant, bee balm, phlox, and monkshood, backed by *Viburnum sargentii* 'Onondaga' berries turning to red. • A Great Dixter–inspired combination of pink mallow, orange daylily, purple veronica, and a yellow mullein, *Verbascum chaixii*.

all around
a barn

It can be a challenge to achieve
long-lasting and pleasing
color combinations in a flower
garden; the intense summer
heat and sun in northeastern
gardens, even in northern
New England, make good
partnerships all too fleeting.

Weathered barn boards make a
sympathetic background for plants.

While I try my best to choose a color palette and position plants to play off one another for dramatic effect, what's more achievable and satisfying over the course of the season is to work with all of a plant's attributes—foliage, habit, size, and character—to create settings where these attributes can be shown off with smart placement and striking combinations.

The intensity of a flower garden's color doesn't want to be extended throughout the whole of a garden. A flower garden's visual excitement does not want to be diluted by repeated displays of perennial color throughout a property. I learned early on that there are a few perennials that can be used to lend the garden a sense of organization. Ellen Shipman used glossy green peony foliage to bring form to herbaceous plantings, and sometimes she would edge a border with small-leaved *Hosta lancifolia*. I've used both those plants to create order, but I came to see there was a much broader and more exciting level of garden interest to be had in expanding my vocabulary of herbaceous and woody plants for foliage effect. If a garden gets this right, it can engender a different sense of excitement, provide a longer season of interest, and can perhaps be less onerous to maintain. I don't know where I learned this, but I suppose it came from listening to more experienced gardeners talk about their experiments with foliage plants, and the satisfaction they took from working with plants for foliage first, before they considered flowers. Rodgersias, hostas, and ferns were some of the first foliage plants to take a place in my garden, especially around the barn and close to the house, and the list of plants I grow primarily for foliage has grown steadily since.

the barn garden

One of the first areas where I worked with foliage was at the foot of the barn facing the house. The texture of the worn boards formed a sympathetic background and foil for plants on all sides of the barn, with weather-beaten, red-stained boards on the north and east sides, and blackened shingles on the south and west. It provided shelter from winter winds and protection from summer sun, as well as storage for household and garden items.

The passageway between the barn and house is narrow, the two probably having once been connected. It now functions as a dramatic entryway to the garden, the tall barn cloaked with a massive climbing hydrangea, followed by an enormous lilac, leading the visitor from the rather unassuming front yard to the fullness of the Flower and Vegetable Gardens in the back.

The passage between the house and barn leads the visitor past an enormous *Syringa vulgaris* 'President Lincoln' to the Flower Garden beyond.

Climbing hydrangea (*Hydrangea anomala* subsp. *petiolaris*) is one of the most impressive vines that can be grown in a northern garden. It is sometimes planted at the base of mature pine trees or other substantial support; Clarence Hay kept his clipped tight against the rustic stone walls of a fountain. These vines are an unforgettable sight clinging to the side of weathered buildings.

I thought climbing hydrangea could mask the crumbling foundation of the barn, and also look handsome covering its side. This plant has the reputation of being slow-growing, but this one established itself rather quickly, rooting along the length of the foundation. Within a few years it covered the entire wall to the barn's peak—it sometimes takes a while for people to realize there is a barn standing behind it. Its twining branches have pried their way between the barn boards into the interior, and it blooms even there, in front of one of the windows. Outside, the solid wall of lacecap flowers is stunning in bloom. These flowers retain their substance and fade rather gently, but their presence is attention-grabbing for weeks.

The investment in this single plant and the time it took to cloak the barn were worth it. Perhaps it may be a little too successful—smaller tendrils crawl along the edges of the lower beds, into mossy carpets, and climb up lilacs and other shrubs. Watching it perform on the barn has encouraged me to use it elsewhere, both as a climbing vine and a rambling groundcover. It can take more sun in northern gardens, but it does best when not subjected to the hottest exposures, and where it's allowed to display its glossy, lush foliage.

THE BARN'S EAST SIDE

On the east side of the barn are three beds that step down following the footings of the barn foundation. These stone-edged beds were probably once foundation walls for a shed, and over the years had become planting beds, but had been stripped of most of their plants before we arrived. A few remnant *Iris* 'Honorabile' and lady ferns had made their way into the mulch. These beds were challenging—the soil is thin and winter winds funnel between the barn and the house, blowing snow off the beds and blasting plants in bare soil. They required careful thought to determine what would thrive under these conditions.

I wanted to experiment with an alpine lawn, and this seemed like a place where I could explore that concept without having to open any new areas. I planted the middle bed with ground-hugging perennials such as thymes, *Phlox subulata*, *Arabis ×sturii*, an assortment of veronicas, and mossy saxifrages. All but some of the veronicas have taken, and to my delight, *Campanula cochlearifolia* seeds itself around. There is also a slightly taller layer of perennials, of which the most successful is the Himalayan maidenhair fern (*Adiantum venustum*) that has run together with a low, small-leaved barrenwort, *Epimedium platypetalum*. This blend increases in size and interest each year. *Saxifraga* 'London Pride' and *Carex* 'The Beatles' hold one edge, while *Hosta* 'Ginkgo Craig' fills a shaded corner.

I have made some missteps in this planting and, to be honest, while the alpine lawn may have been a good starting point, it wasn't the perfect concept for these beds. They are too small, and the stepped stone walls needed plants of greater height. As a late autumn visitor who was seeing the garden for the

Climbing hydrangea has fastened its tendrils to the entire east side of the barn.

first time remarked, "And the point of this planting is what?" I opted instead to expand upon the idea of an alpine lawn in the Rock Garden, where it enjoys a greater expanse, more direct sun, and more success.

The upper bed is planted with pussytoes (*Antennaria dioica*) in both pink- and white-flowering forms, and with *Hosta* 'Halcyon', a medium-sized, blue-leaved cultivar. These plants thrive in difficult conditions, and the silvery tone of the pussytoes and the soft steely blue of the hosta have filled this space for 15 years.

The lowest section sports the boldest foliage and tallest plants. A star magnolia (*Magnolia stellata* 'Royal Star') and a variegated shrubby dogwood (*Cornus alba* 'Ivory Halo') dominate the bed. The substantial foliage of *Rodgersia pinnata* and a yellow-leaved hosta at the foot of 6-foot-tall *Kirengeshoma koreana* provide a backstop. The midsummer yellow flowers of the *Kirengeshoma* always draw attention. Wispy, silvery gray leaves of Roman

In the shallow, dry soil of the lower bed, *Jeffersonia*, ferns, epimediums, and sedges are encouraged to grow in a carpet of thyme, veronica, and ground phlox.

FAR LEFT Stone retaining walls that may have once been a foundation for sheds now serve as planting beds at the base of the barn and climbing hydrangea.

wormwood (*Artemisia pontica*) brush the rodgersias. A few other shrubs and bold perennials have played a role at times: *Sambucus racemosa* 'Sutherland Gold' with its yellow cut-leaf foliage rose out of the hosta and rodgersia foliage for a number of years. In the plant's heyday, the golden-yellow leaves were so pronounced that it carried the rest of the bed, especially with the assistance of golden hosta. *Cornus alba* 'Ivory Halo' now lightens up this corner with its variegated white foliage; it has the advantage of retaining a healthy sheen all summer long, and can take hard pruning to keep it in scale. Scrambling along the top of another low stone retaining wall opposite is a successful long-term collaboration: the dainty foliage of long-blooming *Corydalis lutea* intermingling with the large, cabbage-like leaves of *Bergenia cordifolia*.

Except at midday, these beds receive little direct sun. Their foliage enlivens the passageway with variegated yellow and white contrasting with bolder greens of rodgersias. There is a subtle play of textures that invites lingering

inspection, especially in a transitional area where *Fothergilla* 'Blue Shadow' and *Aralia cordata* 'Sun King' are faced down by *Hydrangea* 'Annie's Blue'. Although plants go in and out of flower, the composition is not dependent on their color. It never seems you've just missed something special, or that a better show is coming along.

THE NORTH FACE

The yellow highlights in the beds are picked up by a group of *Rosa* 'Harison's Yellow' in the adjoining bed on the north side of the barn. Yellow roses intermingle with an unnamed pink rose, and a thicket of snowberry, all of which were in place before my arrival, as was a short stone edging wall and an empty planting bed. I set out a few of the plants I brought from my previous garden in this bed 25 years ago. *Hosta* 'Sieboldiana Elegans', *Amsonia tabernaemontana*, *Rodgersia podophylla*, and *Hosta plantaginea*, along with *Iris pseudacorus* left over from the former garden, are all stalwarts. Their foliage complements one

FAR LEFT Wispy, silvery Roman wormwood (*Artemisia pontica*) pairs well with hosta and rodgersia.

MIDDLE Shaded from afternoon sun, *Kirengeshoma koreana* sports outfacing waxy yellow blooms for most of summer.

LEFT Delicate *Corydalis lutea* mixing with large-leaved *Bergenia cordifolia*.

another, and there is a progression of bloom from May until August. *Rodgersia* emerges in May with utmost vigor, reaching high with its bronze fronds. It follows up with creamy white plumed flowers, just as pale blue stars cover *Amsonia*. The late-summer white trumpets of *Hosta plantaginea* are among the most fragrant flowers in the garden, and with deadheading, the rodgersia's bold foliage commands interest until frost.

Because this part of the bed is narrow, I've planted it with bold-textured plants that act as sturdy companions for one another. Foliage texture and the season-long rhythm of the shapes is more important than any floral interest. The floral highlight of the bed, however, is a month-long bloom of white *Trillium grandiflorum*. It is tucked between a large hosta and the stone wall, and emerges in early May just after the frost is gone from the ground. The sheath that covers the stalk of this plant's inflorescence rests above ground while there are still ice crystals around. As the ground warms, the stalk quickly elongates and begins flowering over the course of a few days. This trillium's pure white funnel-shaped flowers, framed by a whorl of three leaves, come at the high point of spring, and their slow fading from waxy white to pink is

a bittersweet reminder that warmer days are on the way. A few weeks later, I don't notice the plant at all, as the ferns and hostas that are its neighbors have elongated their foliage surrounding it.

The trillium has increased in size, and some years I dig pieces from its sides while it's in flower to move them into a more natural woodland setting. I was also given a double form of this white trillium, and planted it at the far end of the bed. Its flowers are so heavy that it bends under the weight of almost any amount of moisture. It is an interesting form, and I am happy to have it, but the single white trillium is pure joy. The same generous gardener gave me a double form of the native bloodroot as well, and I tucked it under a viburnum, where it has spread nicely.

FAR LEFT *Hosta* 'Sieboldiana Elegans', *Amsonia tabernaemontana*, *Hosta plantaginea*, and *Rodgersia podophylla*, with stray lady ferns, have held this position for more than 25 years.

MIDDLE A long season of interest earns *Rodgersia podophylla* a prominent place in the garden.

LEFT Palmate bronze foliage of *Rodgersia podophylla*.

It appears that after the McKenzies stopped farming, they made more of an effort at gardenmaking behind the barn. A Colorado spruce (*Picea pungens*) was their focal point, probably meant to screen the open back bays of the barn, and when we arrived, the spruce was in perfect condition. It was a classic Christmas tree shape at 35 feet tall and uniformly shaped from ground to tip, holding its lower branches and serving as an effective screen. Now, it has gradually shed its lower limbs, as these spruces nearly always do in the more humid East, and it's become less effective and something of a mixed blessing.

It has also presented an opportunity. In shedding its lower limbs, the spruce has opened access to a consistently moist spot between it and the barn. Although the native soil holds moisture fairly well, there was not a good

LEFT *Trillium grandiflorum* RIGHT Double form of *Trillium grandiflorum* OPPOSITE Tucked between hostas and ferns on the barn's north side, *Trillium grandiflorum* is backed by the emergent bronze foliage of *Rodgersia podophylla*.

moist, shaded spot in the garden that would allow for cultivation of plants that require both moisture and protection from the sun. It is a special spot where any number of desirable Asian woodland plants would be at home, and I began by adding a Japanese wood poppy (*Glaucidium palmatum*).

I had experimented with Himalayan blue poppy (*Meconopsis*) in a couple of shaded parts of the garden, but kept losing it where it didn't get enough moisture. As the spruce shed its lower limbs, I began cultivating a small triangle behind it. For the first few years, the spruce cast sufficient shade from the midday sun that the blue poppies increased in size and began to bloom. I cleared out debris behind the barn and planted a couple of smaller deciduous shrubs to protect the planting from sun, and prepared the bed with generous applications of composted manure and shredded leaves. The bed is now backed with rodgersias to create a neutral green background for the poppies. *Androsace studiosorum* 'Chumbyi' blooms a soft pink at 6 to 8 inches, at the same time as the poppies, and helps to keep inquisitive feet from walking into the poppy bed. In a good year, I will be rewarded with the heavenly blue of poppies in bloom throughout the month of June.

I have focused on *Meconopsis* 'Lingholm', obtained plants from a few different sources, and I've collected the seeds to enlarge the planting. I also grow *M. betonicifolia* (now *M. baileyi*), but find 'Lingholm' a stronger, more perennial plant—and bluer too. The taller stems of 'Lingholm' and its extraordinary sky-blue flowers, with their overlapping petals and prominent

ovary, surrounded by golden stamens—these attributes have justified giving so much space to this hard-to-please plant. Almost as delightful as its stamens are the soft, saffron- and straw-colored bristles on its emerging paddle-shaped foliage. In a good year, when growing conditions are right, each clump will send up strong 3-foot-tall stems. There is nothing like checking each day to observe the progress of this plant's blue flower bud teasing apart its hairy outer sepals. Toward the middle of the flowering season, I may see an emerging bud, a flower in full bloom, and a hairy seed capsule all on one plant.

I still have much to learn about the cultivation of blue poppy. It is entirely cold hardy and responds to the addition of composted manure, garden compost, or shredded leaves, but it's also prone to succumb to summer weather. I have watched large, established clumps melt during extended periods of heat and humidity. I understand that it's when nights don't cool off sufficiently that these plants struggle the most. Although they prefer a rich, moisture-retentive soil, they also need good drainage to prevent root rot, possibly better drainage than I'm able to provide. They failed to bloom one year during an extended wet spring and early humid summer.

FAR LEFT *Glaucidium palmatum*

MIDDLE *Meconopsis* 'Lingholm', in a moist pocket protected from the worst of the scorching sun, blooms through the month of June.

LEFT Sky blue flowers of *Meconopsis* 'Lingholm', with overlapping petals and prominent ovary surrounded by golden stamens.

I have found one other place in the garden where *Meconopsis* are robust and bloom reliably: on the north side of the porch, which gets full morning sun, and where they can sit just beyond the drip line. The soil here doesn't dry out, but excessive moisture drains away.

My fascination with Himalayan blue poppy is due partly to the challenge of growing it well. There is still much to learn, and many a gardener grows it better than I ever will, but I've been spurred on by books and the thought that there are other plants of Himalayan origin that are capable of performing well in my garden.

With so many unfamiliar plants I've tried, it hasn't been easy to know under what conditions they grow in the wild, or how to approximate those conditions in the garden. An example is *Rheum alexandrae*, a small, strapping-leaved rhubarb I got from Bill McNamara of Quarryhill Botanical Garden in Glen Ellen, California, who had collected seeds in China. I have grown this plant for close to 15 years and moved it a few times, each time coming closer to the conditions I think it desires, but it has never grown to the size photographs show it could, nor has it flowered or sent up its ghostly white bracts—the very feature it is grown for.

Rheum tibeticum, a Himalayan rhubarb, took repeated attempts to find just the right conditions.

FAR RIGHT Viburnum plicatum f. tomentosum 'Mariesii'

I knew it wanted moisture, but of what type? I saw photographs of it growing on a river's edge, in colonies on hummocks along meandering mountain streams, in flat meadows, and on rocky, steep hillsides. I surmised it liked readily available moisture, but what role did drainage play, and what other factors had I not taken into account? I doubt mine will ever send up a flower stalk with its showy bracts; nearly every photograph I've seen shows it thriving under heavy cloud cover during the growing season. Apparently it flowers in Scotland and even in the Beth Chatto Gardens in England. I suppose it's possible it is slow-growing, and there may come a summer when it doesn't disappoint. For now, its upright, shiny green, lance-shaped foliage is sufficient.

A friend, knowing I collected ornamental rhubarbs, passed along a seedling *Rheum nobile* he was confident I would succeed with. It was spring and we were in a mild, moist stretch, and there was a gap in the bed next to the blue

poppy in what I thought would be perfect conditions. As I don't have a cold frame or other easy place to acclimate plants with controlled light, shade, and moisture (I am away too much to manage a space that needs daily monitoring), I planted it in well-prepared soil at the base of a rock wall. It doubled in size the first few days in the ground, but the mild stretch gave way to three days of 90°F temperatures with sun and humidity, and it shriveled and declined the longer the heat wave lasted, even with protection from the sun. I'm grateful for the chance to have tried it—and killed it. Horticulturist Dan Hinkley, who saw it in China, didn't think it would stand a chance in the United States, but perhaps the friend of a friend who started the plants from seeds is succeeding.

Another ornamental rhubarb I have experimented with is Tibetan rhubarb (*Rheum tibeticum*), which has leathery, heart-shaped leaves. I have moved it about repeatedly, trying it in moist, rich garden soil and on a shaded slope

where it lived but didn't thrive. After seeing photos of other species rhubarbs growing on rocky slopes in the Himalayas, it dawned on me that if I were willing to sacrifice a healthy clump of cardinal flower that had seeded among rocks in a low wall, it might succeed there. Its new placement allows good drainage, so the crown can stay above standing water, yet the roots remain cool and moist because the rocks act as a mulch to moderate soil temperatures and retain moisture.

Raised slightly above, in a better-drained position, is a double-file viburnum (*Viburnum plicatum* f. *tomentosum* 'Mariesii'). It was an experiment—the best local nursery didn't sell it 20 years ago because it was thought not to be hardy. Its 12-foot span of lacecap flowers on horizontally layered branches adorns the garden in late May and early June. The array of flowers is such that its season is quite extended, with creamy white buds breaking open to reveal pure white. Later on, the pollinated flowers mature into red, berry-like drupes that are eaten by catbirds in August. Its serrated leaves are also doubly arranged opposite one another, on smooth gray branches that ask to be touched. This ruffled foliage turns from reddish purple to deep burgundy

FAR LEFT *Geranium dalmaticum* clothes the ground at the base of the viburnum, with the added benefit of blooming at the same time.

over a long period in fall. It has since experienced temperatures of -28°F and has survived, although its growth was stunted the following summer.

Sweeping underneath the viburnum's spreading branches is *Geranium dalmaticum*, whose glossy dark evergreen foliage flushes out in spring, followed by light pink flowers that coincide with the viburnum's. This is another old-fashioned plant that is useful and takes very little to maintain, but can be hard to find in the nursery trade. It's easy to put a trowel or fork under the rhizomes and move them to fill in other dark, dry, or challenging spots.

Flowering in the shade of the viburnum in early May is double bloodroot (*Sanguinaria canadensis* 'Multiplex'), whose 3-inch whorls of blistering white flowers flutter in the slightest breeze. Its leaves expand later on to play an active role in the foliage theme of this planting.

The moisture-retentive soil in this bed allows for a number of other choice plants, each of which enjoys its flowering season, but whose foliage combines to bring longer-term interest to a plant-rich environment. The bed is home to the mounded shapes of mountain hydrangeas, hellebores, peonies, and hostas, along with coarse-leaved primulas and fine-textured epimediums, and the

strapping upright foliage of sedges and *Iris prismatica*. On the slope above, *Rheum palmatum* var. *tanguticum*, *Hydrangea serrata* 'Kiyosumi', *Gentiana asclepiadea*, *Buxus* 'Glencoe', *Clematis* 'New Love', and *Disporum flavum* thrive in a drier position.

Later in the month, an unnamed pale yellow hellebore, whose softness stands apart from the many vibrant flowers of the spring garden, is accompanied by yellow primulas and a contrasting purple hellebore. These are followed by *Epimedium ×warleyense* 'Orangekönigin' and *Uvularia grandiflora* while the viburnum is in bud.

Sometimes overlooked are 4-foot-tall mahogany blossoms of *Lilium ×martagon* 'Claude Shride' of late June and early July. I also moved this lily around until it found the conditions it wanted, and it has since rewarded me with dozens of densely packed copper-red flowers. These face downward, and bear large golden stamens and upturned petals, speckled with orange highlights. 'Claude Shride' flourished after being moved from a level, sunny location to a slightly sloped woodland edge, protected by the branches of the viburnum. Here it enjoys a full morning of bright light, while the viburnum and nearby birches

LEFT Hellebores, primulas, and *Uvularia* bloom amid the showy foliage of rodgersias, irises, and epimediums. MIDDLE A favorite pale yellow hybrid hellebore. RIGHT Sedges, primulas, and peonies share a moister spot.

protect it from the afternoon sun. It leans toward the light, and I sometimes prop it up with an inconspicuous hooped stake to keep the flower-laden stems from snapping in the rain. Butterflies adorn the flowers for the three weeks they are in bloom.

Summer is animated by multicolored clouds of meadow rue (*Thalictrum rochebruneanum* 'Lavender Mist') in lavender, pink, purple, and yellow. These lead into the more saturated pinks, blues, and reds of late summer. The graceful, nodding bells of the willowleaf gentian (*Gentiana asclepiadea*) overlap with the scarlet of cardinal flower (*Lobelia cardinalis*) and make for too patriotic a combination in August, but the scene calms with the arrival of pink flowers of *Anemone hupehensis* var. *japonica* 'Pamina' in September.

A drier section, also on the north side of the barn, backed by antique shrub roses and a mass of rodgersias, hosts more hellebores, gentians, veronicas, ferns, and *Anthericum*. A new addition to these beds is the long-blooming intersectional peony *Paeonia* 'Garden Treasure'. It took a while to find the

Lilium ×*martagon*
'Claude Shride'
finally settled
in on a slightly
sloping, raised
piece of ground
protected from
hot afternoon sun.

perfect spot for this superb plant. 'Garden Treasure' lives up to its name, and its large, fragrant yellow flower, with dazzling red flares and pink stigmas, earns its place front and center. The American Peony Society recognizes this cultivar as a superior landscape plant because it is sturdy, floriferous, and has long-lasting handsome leaves. I wanted the plant not only for the beauty of its flowers, but also to contrast with the fine-textured foliage of one of my favorite conifers, hiba arborvitae (*Thujopsis dolobrata* 'Nana').

I brought this evergreen back from the West Coast in my carry-on luggage. It's a choice conifer I rarely see offered for sale, perhaps because it is slow growing and always fairly small when I have found it, but it is a refined plant even then. The texture and color of its foliage is most attractive, with deep

FAR LEFT The tall, arching bells of meadow rue lead into the more saturated colors of late summer, with the willowleaf gentian and cardinal flower.

TOP MIDDLE *Anemone hupehensis* var. *japonica* 'Pamina' is a prelude to autumn.

BOTTOM MIDDLE *Paeonia* 'Garden Treasure'

LEFT A long-blooming intersectional peony, *Paeonia* 'Garden Treasure', with sprays of *Anthericum ramosum* and Himalayan blue poppy.

green cedar-like fronds that spread outward evenly around the plant, giving it a neat, symmetrical, almost circular habit. Eventually it will form a layered mound. There is a beautiful example of the mature form at the remarkable Heronswood, a botanical garden in Kingston, Washington, established in 1987 by Dan Hinkley and Robert Jones. It will be some time before mine attains that size, and I'm not eager for it to outgrow its present space—I had no idea what it was capable of when I took a chance on it.

Like a number of other plants I've acquired without knowledge of hardiness, I first planted *Thujopsis* in a protected situation, where it settled in for a few years. Once I realized that it needed more sun than where I had planted

LEFT The feathery foliage of *Thujopsis dolobrata* 'Nana' with diminutive *Aruncus aethusifolius* in bloom, watched over by hosta, rodgersia, and peony. RIGHT A late July-blooming *Anthericum* of uncertain name.

it, I moved it to its present location. The original site is often the last place to thaw out in spring, and in my eagerness to move the hiba arborvitae, I chipped through a couple of inches of ice, carving out a perfectly round "root ball" 4 inches thick that showed a few tufts of roots protruding from the bottom. I scratched out the sides of the new planting hole, leveled its bottom, and set down the chunk of ice with the arborvitae in it. At that point I realized just how hardy this plant is.

A beautiful groundcover plant that helps hold this composition together is European ginger (*Asarum europaeum*). It has been surprisingly easy for me, while some of the other gingers have failed to thrive. Its glossy, evergreen, kidney-shaped leaves are a good contrast to plants of fine foliage or coarser texture. In this case, it ties together the slender leaves of *Amsonia* and *Thujopsis* with the bolder foliage of *Rodgersia podophylla* and hostas. The ginger has begun to seed itself around the garden in the dry thin soil at the foot of a neighboring lilac, as well as in the crevices of broken concrete in the Rock Garden. In its original location, it does not require much moisture; European ginger grown in good garden soil with some shade can go without supplemental water even in the driest seasons. Some gingers under the nearby lilac wilt in the late afternoon sun, but they are increasing in size, and I don't water them.

I am also experimenting with other low-growing, drought-tolerant groundcovers. *Androsace studiosorum* 'Chumbyi' handles the dry, root-filled soil under the spruce quite well. New in the past couple of years is *Vaccinium macrocarpon* 'Hamilton', a slow-growing dwarf evergreen cranberry, which should like the more acidic soil under the spruce. *Anthericum ramosum*, St. Bernard's lily, from central and western Europe, begins its prolific bloom with 'Garden Treasure' peony and continues for weeks, followed by a later-blooming, taller plant that came to me as the species *Anthericum undulatum*, though this isn't entirely clear.

The tranquil planting of the moss bed under the spruce came about by favoring those mosses that established themselves as the spruce shed its lower limbs. Handpicking grasses and weeds that try to insinuate themselves into the bed has helped this along.

Weeding the moss bed is worth it: this has become a gathering place for small groups to view the garden shielded from the sun or in a light rain. It is slightly elevated above the Flower Garden, and offers views across to the fields and hillsides beyond. There are also portals to other parts of the garden from this central point. Although the grassy ground slopes away from the mossy bed, the combined space is an intimate one that draws people to it. It is where we served champagne after our wedding, and where our friend Herb Ferris, an artist, placed a wedge of granite on the moss into which he engraved a gold-leaf band as a wedding gift.

The dominant plant next to the back porch is the 'President Lincoln' lilac that the McKenzies planted, probably in the 1920s. It was introduced in 1916 in Rochester, New York, a center of lilac hybridization. Its fragrant lavender-blue flowers bloom mid-season; the rest of the year it stands out for the thick sinewy trunks that reveal its venerable age. A half-dozen main stems support flowering branches 20 feet wide and tall. One of the advantages of this variety is that it doesn't send out suckers from the base. The lower branches spread nearly horizontally and frame views into the garden.

The soil beneath the lilac does not wish to support anything much more than mosses and the native lady ferns that have put themselves there. The southern and western sunny sides are treated the same as under the spruce—they are weeded to keep the grass out, but self-sowers from garden plants are allowed—a few European gingers, as well as a smattering of helle-bores and cyclamen.

The north, shady side under the lilac is treated more intentionally. There is a swath of hellebores that have taken to these conditions. Nancy Goodwin of Montrose, a historic garden in North Carolina, encouraged me to experiment with hellebores, starting with the hardy Christmas rose (*Helleborus niger*), native to mountainous regions of Italy and central Europe. Her recommenda-tion came shortly before a trip to the Ligurian Coast south of Genoa, where, on a hike on a cold December day, we came across masses of Christmas roses in bloom. Nancy's display at Montrose is almost as impressive, and maybe more so, because she brought together a few different strains of the plant to compare and contrast under the shade of a dawn redwood.

I started with a few plants of the straight species *Helleborus niger*, and as I came across other strains I added to the collection, especially early-blooming *H. niger* 'Praecox', *H. niger* 'Potter's Wheel', and *H. niger.* ssp. *macranthus*, which I purchased from Plant Delights Nursery, also in North Carolina. The last is a bit taller than the others, with larger flowers and darker green foli-age. There has never been any question about the hardiness or vigor of *H. niger*—the half-dozen plants I began with have flowered, produced seeds, and increased in number. Some years, especially during a mild December, there can be a fair display of their cup-shaped white flowers before Christmas, blooming through the snow. Other years they make a more extravagant show-ing as soon as the snow recedes and days warm up, often keeping it up into the later part of May.

As much as I'm grateful for what can be the first substantial display of flowers in spring, it is these plants' leathery, dark green foliage that holds my interest. It needs minding in fall to remove fallen leaves, and in spring, prun-ing to remove blackened foliage, but that's a small price to pay for a handsome clump of plants that stays healthy and robust all season long. I've combined them with a delicate, low-growing willow, *Salix brachycarpa*, to complement and contrast the foliage of two plants that prosper in similar conditions.

Christmas rose blooms among low-growing willows on the mossy bank beneath *Syringa vulgaris* 'President Lincoln'.

THE WEST SIDE

The west side of the barn began as a setting for a collection of New England asters. Given that my first efforts at ornamental horticulture were with the flower gardens in Cornish, some of my initial collecting concentrated on old-fashioned perennials. Most of what I collected were unnamed varieties of asters gathered from those gardens. I was inspired to plant them as a group after seeing my friend Virginia Colby's lively planting of pink-, purple-, rose-, and white-blooming asters at her Cornish farm. Virginia scanned the road-sides for naturally occurring color variations, and when she saw one she liked, she would dig a piece of it for her plot. Not long after this, I saw an incredible display of New England and New York asters at the Bagatelle in Paris.

A true find was an aster growing at Aspet likely introduced by Ellen Shipman. This plant grows to about 4 feet, with dark green foliage, never suffers from heat or drought, and has cheerful purple rays with a yellow center. I haven't discovered its name, but it is definitely an improved selection of New York aster (*Symphyotrichum novi-belgii*). It's an exceptional plant that I have

FAR LEFT *Hydrangea paniculata* 'Tardiva' separates the lusher plantings on the north side of the barn from the quieter, more shade-tolerant border on its west.

MIDDLE *Vancouveria hexandra*

LEFT *Epimedium grandiflorum* 'Lilafee'

divided and used in other parts of the garden, and shared with other gardeners and nurseries. In addition to collecting from gardens, I've also purchased new varieties, especially those from German hybridizers. A favorite is the deep rose *Symphyotrichum novae-angliae* 'Andenken an Alma Potschke'.

After a while I realized that asters were not particularly well suited to the dry soil by the barn. The bed was lacking in early-season interest, and I could take better advantage of the weathered barn board as a setting. A pagoda dogwood (*Cornus alternifolia*) seeded itself into the lower end of the bed, and I added a couple of meadow rues and cardinal flowers. Quite by accident, this inspired combination leads into aster time. The purple fruits of the dogwood along with its red pedicels pick up the lingering purples of *Thalictrum rochebruneanum* 'Lavender Mist' and the scarlet of the cardinal flower (*Lobelia cardinalis*). But eventually the asters failed because the soil was too dry, and they lost their lower foliage. I moved them and began rethinking the bed.

What's now in place is far more interesting and sustainable. The barn was once one side of a three-sided yard. The barns that formed the other sides are now gone, but I've enhanced the sense of its being a yard with the addition of

LEFT *Actaea racemosa* enjoys a long season of bloom, opening from bottom to top, with small, round seed heads nearly as showy as its flowers. RIGHT A sculpture by Herb Ferris, *Bill's Arrow*, emerges out of a mass of Russian cypress and Himalayan rhubarb.

a number of paper birches that create a canopy and sense of enclosure. The 40-foot border begins at the southwest corner of the barn, with a clump of paper birch alongside a panicle hydrangea and *Helianthus* 'Lemon Queen'. The northwest corner of the barn is anchored by an enormous *Hydrangea paniculata* 'Tardiva'. The planting extends another 20 feet beyond to the doublefile viburnum.

I decided I would avoid any watering after the plants became established, and I've chosen shrubs and hefty clumps of perennials that hold the entirety together. The perennials at the drier end are mostly small-leaved drought- and shade-tolerant groundcovers like *Epimedium grandiflorum* 'Lilafee', *Vancouveria hexandra*, and *Geranium dalmaticum*, along with the thick-stemmed wood fern *Dryopteris crassirhizoma*. The soil improves in the midsection, where tall and large-leaved perennials and shrubs can thrive, including *Kirengeshoma palmata*, *Lamium orvala*, *Deinanthe caerulea*, *Hosta* 'Sum and Substance', *Actaea racemosa*, and assorted hellebores.

There is some floral interest here, which gets underway in May with mauve pink pouches of *Lamium orvala*, a bonus plant totally unknown to me that I received in an order from Far Reaches Farm, a nursery in Washington state. *Epimedium* and *Vancouveria* are both extremely subtle in flower, but always a delight. *Deinanthe caerulea* is a hydrangea relative, and its nodding blue flowers invite close inspection in late July.

In the bed are boxwoods that are an experiment in plant hardiness—*Buxus* 'Glencoe' (syn. 'Chicagoland Green'), somewhat rounded in shape and quite similar to *Buxus* 'Green Velvet'. In its early years here, it showed no dieback even after the most severe winters, but lately those exposed to more winter sun have shown a bit of dieback, which is easily pruned out in spring.

Native black snakeroot (*Actaea racemosa*), a favorite of Cornish gardeners, throws up a dozen white spires to 8 feet in mid-July. Its oddly fragrant flowers are covered with pollinators—honeybees, bumblebees, and hoverflies—for as long as they're in bloom. The plant's seed capsules are nearly as showy, extending interest well into August, by which time neighboring *Kirengeshoma palmata* has taken the stage with its pendulous flowers in waxy yellow. The tall, mounded foliage of each gives the garden substance through the season.

The late season sees another bugbane sending up multiple spires with close-cropped white flowers in *Actaea japonica* 'Cheju-Do', hailing from Korea's Cheju Island. Finally, October brings the eccentric, light pink, brush-like flowers of Japanese shrub mint (*Leucosceptrum stellipilum*), protruding from the top of handsome foliage. This always feels like the last gasp of a season fast winding down.

Although I'm mainly after a long-lasting and relatively maintenance-free foliage composition, there is almost always something in flower in this bed from May to October; the strategy is for there to be a modest display of floral interest staged throughout the season. I don't care to have a large show, but rather my aim is more about enjoying the foliage ensemble with one or two choice plants in flower at a time. This is a strategy I witnessed to great effect at Peckerwood Garden, an incredibly plant-rich public garden in Hempstead, Texas, outside Houston, originally made by artist and teacher John Fairey.

There is a break in the plantings beyond the viburnum, with a grass path leading up to the birch lawn; on the other side of the break, set atop a stone wall, is another sculpture by Herb Ferris. *Bill's Arrow* is a composition made with amphibolite, a striated metamorphic rock of gray and black. The sculpture's point is directed skyward, its shaft a railroad tie positioned on a granite base that rises out of a mass of Russian cypress. We've shown other pieces of Herb's figurative work in the garden at various points, but this is my favorite—it is the focal point I was looking for on the wall. Its natural materials, slightly worked by Herb's hand, are an earthy and spiritual abstraction.

the dining room

Though its physical relationship to the Barn Garden may seem tenuous, I connect the small garden outside our dining room and kitchen with the barn and its plantings. Like those, it's associated with a structure—the house, in this case—and is composed to give year-round satisfaction when viewed from within the house or from the garden.

The view from the dining room window is the most glorious inside-outside view in the house. For 25 years, in all four seasons, the knotty arms of a 'Montmorency' sour cherry tree stretched across the windows, inscribing a view of the landscape beyond. In May, it was a snowdrift of white flowers, which, when they dropped onto flowering blue *Vinca minor* below, intimated a scene out of a Japanese garden. The tree possessed a character that only comes with age, yet age claimed it. For some time we witnessed its slow decline until, one humid summer, it lost all its leaves to a fungal infection, and the following spring failed to send out any new growth. Its sinewy trunks and lateral branches were a living sculpture in winter; in spring, every branch was covered with blossoms alive with honeybees. We picked its red, mouth-puckering sour cherries for pies or sauce if we could beat the bluebirds, thrushes, catbirds, and waxwings to them.

Its loss was doubly felt, from inside and from outside, especially since the planting of the slope below was organized around it. The McKenzies understood the ornamental value of this hardy, old-fashioned tree. The slope under it is retained with a massive granite stone wall into which steps and flagstones leading to a back door and summer kitchen are set.

On the slope surrounding the sour cherry, four *Picea abies* 'Pumila' create mounds that hold the bank and create a grounded but not too solemn presence. To keep it lively, I placed a female winterberry (*Ilex verticillata*) between them and the kitchen window. I learned from observing the sour cherry how pleasing it can be to gaze out of a window through a deciduous tree or shrub, so long as the leaves are fine-textured and the branching pattern mostly horizontal, with the branches pruned to allow views between. The spruces offer a solid background and ground it visually. The winterberry's fine-textured foliage remains healthy in summer, and its branches are covered in red berries in fall and early winter. Winterberry is inconspicuous in flower—its primary ornamental value

Mounds of bird's nest spruce, smooth hydrangea, and spirea
hold the bank below the kitchen and dining room.

Old-fashioned hills of snow hydrangea.

FAR RIGHT Pleasing from inside on a melancholy fall day: tawny heads of smooth hydrangea, dark green of spruce, epimedium, and *Amsonia*, with a hint of winterberry.

is in its berries—but as with the sour cherry, the structure of the plant is its most interesting feature. It brings the birds closer to the kitchen window too. Dutchman's pipe (*Aristolochia durior*, syn. *A. macrophylla*) clothes the end of the deck, and its large rounded leaves are an appropriate coupling with the stiff texture of the spruces.

On the far side is a bridal wreath spirea (*Spiraea* ×*vanhouttei*) from the McKenzies' time. I've brought it into the larger composition with a low mat of *Epimedium sulphureum*, a good-size chunk of *Amsonia tabernaemontana*, and another *Picea abies* 'Pumila'. Periwinkle (*Vinca minor*) ties it all together.

The lichen-covered granite capstones that hold the bank back create a pocket that is filled with hills of snow hydrangea (*Hydrangea arborescens* 'Grandiflora') dug from the bank across the road. The broad, egg-shaped

foliage of this variety of smooth hydrangea makes for a dense mass, while its large (but not too large) rounded flower heads cascade over the walls. This is an older and, I think, superior form of the ever-present *Hydrangea arborescens* 'Annabelle'. Its flowers are more graceful, and it doesn't flop over as readily as 'Annabelle'. Flowering begins in June and fades to tawny brown by late summer, but to my eye, these hydrangeas look their best on a melancholy fall day as counterpoint to the solid granite and spruces, and flecks of red from the winterberry.

As well composed and even beautiful as this and the Barn Garden may be, they exhibit a certain restraint that felt important in their making. My more adventurous forays into composition and abstraction with plants took shape farther away from the house, and farther into the open landscape beyond.

ruined foundations

With its cracked cement floor and crumbling rock walls, the foundation of a former dairy barn on our property suggested a place for a rock garden, and I obliged.

Tiarella, ferns, and primulas, survivors of my initial plantings, took to the shady end of the Rock Garden, while ground phlox and other sun-loving plants spread through the sunny north end.

Birches and poplars rise out of the former barnyard, creating a foyer for the Rock Garden. OPPOSITE We cleared the rubble from the former dairy and stable and began planting.

I planted out the floor with mat-forming alpine plants and expanded the perimeter with a gravel garden and troughs for alpines.

Next to the dairy is a sunken garden made in the remnants of a collapsed stable I've filled with plants that evoke the granite outcroppings of New Hampshire's White Mountains. The sunken garden is adjoined by a series of beds planted under century-old apple trees and a grove of paper birches. A second rock garden area, with better drainage and morning sun, was opened up on a north slope below the deck. Trial and error has determined which plants succeed in these varied situations. Most of the rock garden plants tend not to be true alpines, mainly because the barn foundation turned out to be not especially well drained, and because parts of it are heavily shaded. Successful plants there include willow, primula, gentian, saxifrage, blue poppy, phlox, and a variety of dwarf groundcovers. At home in the Stable are New England natives, low-growing conifers, cranberries, willows, heathers, and grasses. Under the apples and birches are primulas, epimediums, trilliums, ferns, and sedges.

making the rock garden

Four barns once enclosed the yard for a herd, though three are now long gone: a hay barn, a milking parlor with stanchions for 16 cows, and a square stable at the western end. One remaining barn still stands to the east. This grassy rectangle was shaded on the roadside by a pair of wild apples, but today I've made it a grove of poplars and paper birch.

The general outlines of the barnyard only became clear as debris from collapsed roofs and scattered contents was pried apart and carted away. It took a couple of years of occasional workdays to disentangle the tin roofing from the ground. With the roofing gone, we dragged off the sheathing and beams and exposed the concrete floor and granite walls. Bits and pieces of chain, wire, and other rubble were fished out, some of it saved for later use. Gradually, I began to get the sense that these abandoned structures could find new life as gardens, and the excitement of gardening against agricultural ruins took hold.

It seemed the cracked concrete floors would provide good drainage for alpine plants that could work their way under the broken concrete for a cool root run. The soil under the milking parlor was slightly limey from decomposing concrete, and still rich from many years of barnyard manure.

I chose to begin with a rock garden in the milking parlor's broken concrete floor. The former stable next to it was enclosed in lichen-encrusted granite walls, and suggested a place of refuge with simpler plantings. The long

foundation walls of the hay barn could then delineate a lawn with a grove of poplars and birches.

This site was a working hillside farm, and there was nothing particularly extraordinary about its remnants, but these former outbuildings did come with stellar views to the fields and woods beyond. It seemed reasonable to play off the romantic notion of a garden in ruins, and to stabilize the foundation walls and plant in a way that showed them to advantage. There's something of a tradition in Vermont of 20th-century gardeners making gardens amid the ruins of disused barns and farms.

LESSONS FROM THE FELLS

This romantic notion of gardening in ruins was complemented by the practical need to learn more about rock gardening and alpine plants as part of my work. I had recently become the gardener at The Fells in New Hampshire, and was responsible for the restoration of its rock garden. Part of the attraction of working there had been the chance to learn a new range of plants and garden types not present in the Cornish gardens I'd worked in up to that point. For more than 50 years at The Fells, Clarence Hay created perennial and rose gardens, woodland gardens, and rock gardens, and its centerpiece was a rock garden that spanned almost an acre, much in need of restoration.

What made the rock garden at The Fells exceptional was how Hay created a variety of conditions for growing a wide range of alpine plants, yet at the same time made it seem as if the garden were a natural outgrowth of the place. From an abandoned New Hampshire hayfield, he shaped a plausible alpine scene that felt as if it had always been there, not imposed by a plant collector showing off his rarities. There was much for me to learn in working in Hay's rock garden—as much the sensation of spending time in a garden that's part of a natural landscape as how to grow alpine plants, with their exacting cultural requirements. There is a steep learning curve to gardening with alpines, and I approached the renovation of this rock garden slowly and deliberately. Hay grew over 400 different taxa; I was familiar with only a small number of these, and it made sense to learn something about rock gardening in my own garden in the process.

Clarence Hay's rock garden at The Fells appeared so natural that one visitor remarked how lucky Hay was to have a stream running down its middle.

LEFT The Rock Garden's center is a tapestry of phlox, thyme, sedum, and juniper that knits together in the former gutters and spills along the cracks and margins of the foundation. **RIGHT** Self-sowers like pasque flower and mullein are not only allowed but encouraged to sprout within the tapestry.

I realized the importance of creating a variety of conditions for the cultivation of rock garden plants: sunny, hot locations for some, cooler bright conditions for others, overhead shade for woodland perennials and ferns. It took longer to realize that I would not be able to replicate situations that the choicest of true alpine plants require—sunny but cool daytime temperatures, protection from excessive heat and humidity, and perfect drainage.

But I could see that the ruined foundation floors and walls of the barns could be treated as a series of garden rooms with an open rock garden at the core. The milking parlor's concrete slab was fitting for hardy, sun-loving rock plants. Two old apple trees offered shade for woodland plantings, and the stable could become a green garden with low-growing conifers and alpine willows. Eventually, a north-facing gravel bed with plants mostly of Himalayan origin presented a better environment for growing alpine plants.

I observed plants at The Fells that had survived for decades with minimal maintenance: *Geranium dalmaticum* and *G. renardii*, *Sedum kamtschaticum* and *S. sieboldii*, *Phlox subulata*, *Dianthus*, *Veronica spicata* 'Nana', and various thymes still covered a lot of ground. *Lobelia cardinalis*, *Gentiana septemfida*, and *G. asclepiadea* had seeded into generous-sized colonies. Some of these would become indispensable plants here.

The Rock Garden introduces a bit of wildness into the overall garden, in counterpoint to the orderliness of the Vegetable Garden and the Flower Garden's stage-managed complexity. There is the difference in scale but also a sense of freedom in how it contrasts with the others. It's more intimate and feels more in tune with nature. The more relaxed, natural setting of the Rock Garden is a sanctuary where I go to sit or weed and be still in its presence.

Rock gardening is also an opportunity to experiment and marvel at how plants grow. Discovering what conditions plants need to thrive and how to provide them is one of its great joys. A number of plants succeeded from the beginning and became mainstays, through their ability either to cover concrete or to seed themselves into the rubble. I began with a limited number of plants and paid close attention in the early years to which ones took root. Those hardy, ground-hugging plants that succeeded are now the unifying elements of the Rock Garden, foundational plants that tie it together.

SUCCESSES AND FAILURES

Before the Rock Garden was established, two large poplars and an elm had grown through and cracked the concrete floor, and cast cool, dappled shade over the foundations. My first plantings were of shade plants, primarily *Tiarella*, primulas, and ferns. Horticultural varieties of lady ferns and woods ferns did exceptionally well in those conditions; however, this planting didn't last long. First, two large poplars at the south end suddenly collapsed, and then the 30-year-old elm succumbed to Dutch elm disease. In the space of a couple of years, the garden went from pleasant overhead dappled shade to full sun. *Tiarella*, primulas, and ferns would have to grow elsewhere.

After a few years and considered responses to these changing conditions, the garden has finally achieved its own distinctive character. The crevices in the broken concrete, along with the cool, rich soil under it, are ideal for alpines that have knit together into a tapestry. A pea stone edge has been added for individual specimens, the shaded south end under the apples is filled with choice ferns and shade-loving perennials, while a group of hypertufa troughs rests along the edges of the concrete floor. The Stable is an evergreen refuge, with saxifrages tucked into its granite walls.

STANCHIONS AND GUTTERS

The center of the garden is a tapestry of fine-textured plants that shine in multicolored hues in spring and early summer. Ground phlox, thyme, juniper, sedum, saxifrage, broom, pussytoes, and *Dianthus* weave together covering the concrete floor.

I didn't know at first which of these plants would be most successful. I knew that I wanted the plants in the larger open space of the foundation to knit together into an alpine lawn, growing into a community. Phloxes and thymes tie the whole together, and when they're in flower (phloxes in May and thymes in June), the effect is colorfully impressionistic. More valuable still, their foliage persists throughout the year—muted green in summer and contrasting silver, gray, green, and brown in fall and winter, creating a year-round effect.

Thyme has covered larger slabs of concrete, primarily *Thymus praecox* 'Doretta Klaber', along with wooly thyme (*Thymus pseudolanuginosus*). Named varieties of *Phlox subulata* make up most of the ground phloxes, along with a few hybrids and close relatives. These bloom in pinks, lavenders, purples, blues, and whites, and have self-sown into the tiniest cracks in the concrete.

Patched into the alpine lawn are taller growers that I've encouraged to seed around, adding a dimension of height as well as motion, making for a more dynamic composition. Pasque flowers (*Pulsatilla vulgaris*) enliven the scene with purple and red blooms while the phloxes are flowering, and then carry on with the delicacy of muted seed heads when the thyme is in color. June is also the month for fresh white flowers of *Saxifraga paniculata*, flowering at 1½ feet. The saxifrages push up numerous stems, each capped with an array of small white flowers that sway as a group in the breeze.

A narrow gutter, placed at what had been the cows' business end, runs the length of the foundation, and in it grow rock plants that can withstand baking

CLOCKWISE FROM TOP LEFT Pasque flower in purple and red is an easy-to-grow and welcome early-season bloom. • In easily reached sections in the gutter, space is given to individual plants and planting combinations, such as *Arabis ×sturii*, *Gentiana acaulis*, and *Phlox subulata* 'Fort Hill'. • Pale yellow new foliage of *Juniperus* 'Gold Cone' echoes bright yellows of *Genista lydia* in a sea of multicolored ground phlox. All are evergreen, making their presence felt in other seasons. • By the end of May, ground phlox has made its presence known, almost to a fault. Named varieties predominate, but seedlings appear all the time, sprouting out of cracks of the foundations. • *Thymus praecox* 'Doretta Klaber' is among the tightest-growing and most weed-smothering of the thymes. It makes a long-lasting combination with *Sedum* 'Weihenstephaner Gold', happily spreading over the decomposing stump of an elm.

Maidenhair fern and *Luzula nivea*, bookended by silver-leaved *Salix helvetica*, enjoy the coolness of shade.

OPPOSITE Willowleaf gentian, a keepsake from The Fells, introduces a bit of wildness.

heat and don't require perfect drainage. Before I realized the gutters didn't drain well, I lost many delicate growers here. Occasionally a strong patch of gentian or veronica would disappear after a particularly unfavorable winter, or wither in summer heat and drought, but there was always phlox or *Potentilla* ready to move into its place.

In the north end of the garden, where the foundations are more broken up and in danger of collapse, the plantings are simpler, with spreading junipers and masses of ground phlox, thyme, and sedum. The May show of ground phloxes is followed by the June show of the thymes and *Sedum* 'Weihenstephaner Gold'. The combination of the two is so winning that I'm afraid I've let both aggressive plants become rather too expansive. But they cover lots of ground and require almost no maintenance.

Other gaps afford space for willow, gentian, saxifrage, primula, and a large *Juniperus communis* 'Gold Cone'. The decomposing stumps of the elm and poplars form a perfect seedbed for *Sempervivum* and other low-growing, self-seeding plants resistant to drought, including *Verbascum*, *Euphorbia*, sedums, ferns, and mosses. A patch of *Saxifraga paniculata* has steadily increased in the shade of a mossy granite slab.

A southern nook shelters three large troughs in dappled light and a *Fothergilla major*, with epimediums, ferns, *Antennaria*, and *Gentiana*

asclepiadea growing at its base. In deeper shade closer to the apple, there is a collection of troughs with climbing hydrangea as a groundcover to knit them together, along with self-sown gentians and ferns.

Other, taller flowering plants spread color throughout the garden after the groundcovers have subsided, chief among them mulleins in early summer, a few alpine betonies, and cardinal flower in late July through August, followed by willowleaf gentian. Also showing some height in early summer are *Dianthus* and *Paradisea liliastrum* (syn. *Anthericum liliastrum*). The shadier end is anchored with robust clumps of *Luzula sylvatica* and *L. nivea*, and maidenhair fern.

PEA STONE EDGE

Running the length of the foundation on the barnyard side is a 3-foot-wide pea stone–mulched bed that is meant for choice individual specimens. I added it because the plantings on the milking parlor foundation began to feel too small for the space. It has also been a good nursery bed in which to grow plants to a size where they can be introduced into larger beds. I stuck hardwood cuttings of *Salix nakamurana* var. *yezoalpina* into the cool soil in the nooks of the concrete slab foundation so their thick green foliage would mask the shoddy construction. In recent years, I've added a number of epimediums, especially

FROM LEFT In late summer, foliage becomes paramount, but at The Fells I learned of worthy late-season flowering plants for the Rock Garden like this diminutive betony, *Stachys spathulata*. · A cool December carpet of *Juniperus procumbens* 'Nana', *Genista*, ground phloxes, and thymes. · The succulent foliage of *Primula marginata* 'Kesselring's Variety' is equally at home in pea stone and troughs. · An unexpected combination of *Anemonella thalictroides* 'Shoaf's Double Pink' and *Epimedium ×warleyense* 'Orangekönigin' creates buzz. · *Narcissus* 'Thalia' blooms with ground phlox in the low-maintenance north end of this area.

types that spread by rhizomes; these also help to mask the wall. Part of the 40-foot run is gently sloping, but most of it is level.

The pea stone bed has a richer assortment of alpines that want better drainage and less crowded conditions. It's planted with mostly evergreen perennials and low-growing shrubs that are repeated along its length to tie it all together. It is the only portion of the Rock Garden where I've introduced bulbs, such as *Crocus tomassinianus* and *Iris* 'Harmony' and 'Katharine Hodgkin', along with a couple of diminutive narcissus. Low-growing boxwoods, dwarf willows, birches, and azaleas give the grouping height and substance. There is something in bloom here from April to October.

I'm not particularly trying to create flowering combinations in this bed, although occasionally a pairing I would never have thought of shows up and steals the show—most memorably the pink and orange combination of *Anemonella thalictroides* 'Shoaf's Double Pink' with *Epimedium ×warleyense* 'Orangekönigin'. Gentians, campanulas, saxifrages, *Aubrietia*, and *Paradisea* also do well.

A dozen hypertufa troughs are set along the margins of the Rock Garden in sun and shade.

Cyclamens like the pea stone as well, with *Cyclamen purpurascens* and *Cyclamen hederifolium* being the hardiest. In the cooler, shadier end of the bed, primulas thrive, especially *Primula marginata* and *P. pubescens*, along with the delicate *P. farinosa*. The pea stone mulch encourages self-seeding, and some of the campanulas and phloxes have made themselves at home. *Campanula cochlearifolia* runs rampant in some years, along with the taller, hardier harebell *Campanula rotundifolia*. *Allium thunbergii* 'Ozawa' and *Saxifraga fortunei* wind down the flowering season in October.

ALPINE TROUGHS

There are many desirable alpine plants that will grow on the flat surface of the Rock Garden, but there are also those that are more successfully grown in the sharply draining soil of troughs, where they can be arranged for closer inspection. I've placed a dozen hypertufa troughs around the perimeter of the Rock Garden, and these are outfitted with conifers, alpine willows, saxifrages, primulas, and other dwarf alpines that thrive in a freer-draining soil mix.

Most of the troughs come from Stonecrop Gardens in New York, and they are extremely well cast and durable. A cast-iron sink with a drainage hole also makes for a good growing container. The majority of the troughs are arranged so there is a rock or group of rocks placed to hint at a natural setting. Some are positioned to resemble alpine peaks, others to leave crevices for saxifrages to fill. The most interesting is a piece of tourmaline given to me by arborist friend Bill Murphy. This semi-precious coal-black stone is the centerpiece for choice saxifrages and *Primula marginata*.

Most of the troughs are planted with a single dwarf conifer to reinforce their natural appearance and provide year-round interest. Troughs on the sunnier end of the garden are planted with dwarf spruces, junipers, and pines, such as *Juniperus communis* 'Berkshire' and *Pinus mugo* 'Short Needle'. Those on the shady end are planted with dwarf *Chamaecyparis*, hemlocks, and firs.

My favorite trough plants are cultivars of *Saxifraga paniculata*, especially 'Minutifolia', which fills in crevices and covers the pea stone–mulched surface. Its finely encrusted margins sparkle in the crisp air, and there is rarely anything to detract from the beauty of its rosettes. *Saxifraga paniculata* 'Whitehill' also does quite well. *Primula marginata* is another favorite, with succulent evergreen foliage and violet-blue flowers in May.

Dianthus, campanulas, and midsummer-blooming alliums also have a place. *Draba bryoides* var. *pygmaea* gains in diameter every year, and is

LEFT Troughs are planted with a variety of conifers, saxifrages, and primulas in conjunction with found pieces of granite and tourmaline, evoking miniature alpine scenes. **RIGHT** Selections of *Saxifraga paniculata* do particularly well in the troughs, especially 'Minutifolia'.

blanketed in the smallest yellow flowers in May. One trough is covered with low-growing *Thalictrum kiusianum*, always decimated by slugs when grown in the ground, but now covered with long-lasting, delicate meadow rue flowers in July and August.

Plants in the shady end include *Saxifraga pennsylvanica*, dwarf hostas and coral bells, and low-growing sedges and ferns. Most of the alpines, including saxifrage, thyme, and *Dianthus*, are evergreen, so there is always something to appreciate. During flowering season, the saxifrages throw up 10- to 18-inch flower stems, while the flowers of *Dianthus*, *Potentilla*, and primulas hug the plants.

Plants that find the conditions suitable are long-lived and can withstand winter lows and summer high temperatures without showing too much stress. Because they are in shade half the day, they do not require frequent watering except in dry stretches. I pinch back or prune the conifers to keep them in scale and retain a balance between foliage and root mass. The soil needs occasional refreshing, and in that case I lift the plants, tease them apart, add fresh soil and grit, reset the rocks, and then patch in the alpines. Occasional losses are easily replaced by adding new soil and plants selected from the nursery bed. For winter protection, I cover the troughs with evergreen boughs to shield them from winter sun and deer browse. The troughs themselves are extremely well made and have not spalled or cracked.

HIMALAYAN GARDEN

Eventually, I realized there was one other place in the garden where I could experiment with alpine plants—and it turned out to be more suited to alpines than the original Rock Garden. The slope below the deck and porch faces north, and sees full sun from dawn until noon, and cool shade the remainder of the day, and since it collects runoff from the roof, it rarely dries out. The ground slopes just enough so that there is never any standing water, and because it is so close to the north side of the building it remains snow-covered all winter, sometimes into May. It much more closely resembles true alpine conditions than the floor of the milking parlor. For a while after I realized the conditions here were suitable for alpines, I held off creating a bed, because I worried it might seem unconnected to the rest of the garden.

Then two books came to my attention, and the idea struck to make a garden featuring Himalayan plants. The *Guide to the Flowers of Western China*, by Christopher Grey-Wilson and Phillip Cribb, spent months close at hand on the coffee table. I flipped through this volume for hours, its two thousand color photographs depicting gentians, peonies, clematis, blue poppy, and dozens of other genera I had never heard of, and with flowers unlike anything I had encountered before. I knew the vast majority was beyond reach for growing in this country, but an interest took hold nonetheless.

Jim Jermyn's *The Himalayan Garden: Growing Plants from the Roof of the World* demonstrates that some of these plants could be grown in garden settings, and so I was motivated to pursue them with enthusiasm. I began by digging up a 4-foot-wide piece of sod the length of the porch. I laid down a few flat stones on the surface—I thought that trying to create a rockery with protruding stones might look too contrived this close to the building.

I made no attempt to evoke nature in this part of the garden. Any attempt to do so would ring false, given its proximity to the lattice skirt under the

porch, relatively small scale, and contrived rectangular layout. I would have preferred a larger and more discrete area, but that was not available. The bed looks more like a curated collection than an alpine scene, which is fine.

Into the native soil I mixed turkey grit, peat moss, sand, and compost. I knew that the native soil would retain moisture, and that the plants that would thrive in it would be those that appreciated consistent moisture rather than rapid drainage. I've accommodated a few of those by removing soil and adding extra grit to the mix, but the plants that grow best are primarily those that tolerate or enjoy consistent moisture. The surface is pea stone mulch, which makes for ease of weeding and is easily replenished when disturbed.

The initial layout was mainly with plants native to the Himalayas I acquired from specialty nurseries. These included: *Saxifraga lilacina*, *S. oppositifolia*, and *S. decussata*; along with epimediums such as *Epimedium wushanense*, *E. fargesii*, *E. latisepalum*, and *E. lishihchenii*; and a number of Himalayan primulas like *Primula chungensis*, *P. florindae*, *P. alpicola*, and *P. bulleyana*. The straight species *Corydalis elata* has thrived, while newly introduced cultivars of Chinese origin, such as *C.* 'Blue Heron', have not. Himalayan

LEFT On a gentle, north-facing grade just below the deck, I cut a 6-foot-wide bed into the turf. Grit and pea stone were added to promote drainage, and a variety of Himalayan alpines, including *Primula capitata*, were planted. MIDDLE Lower-growing saxifrages, primulas, campanulas, *Androsace*, and *Dianthus* spill the length of the bed, while taller epimediums, ferns, *Corydalis*, and *Meconopsis* hold the back. RIGHT The apricot blooms of *Saxifraga* ×*megaseaeflora* 'Galaxie' are held just above its dense cushion of silvery leaves.

blue poppy also does well here, taking to the bright morning light and consistently moist soil.

The Himalayan conceit didn't last all that long, as I began to fill the bed with other hybrid saxifrages and choice plants such as the European *Gentiana acaulis* that struggled in the original Rock Garden, but thrived in this more open, sloping site. The pea stone mulch is also a seedbed, with primulas in particular making themselves happy in mossy parts. I have twice enlarged this garden by stripping away a parallel length of sod and filling in with additional pea stone.

The flowering season in the Himalayan Garden begins with *Saxifraga sancta* var. *macedonica* blooming as the snow melts, its compact evergreen hummock revealed and covered with cheery yellow flowers. This is one of the few Kabschia saxifrages I can grow. It's the earliest to bloom, but more difficult in the open garden than some of the later-flowering *S. paniculata* types that also populate the garden. A second Kabschia type is *S. ×megaseaeflora* 'Galaxie', which blooms a peachy yellow. It is pink in bud when it emerges from its silvery green cushion, and when spring is cool, the flower is a shade of apricot orange as well.

There are a number of Kabschia saxifrages I haven't succeeded with—as true alpine plants, they want to be snowbound throughout winter, and in summer bathe in bright thin air and mountain breezes without a trace of humidity. If a gardener can come close to providing those conditions, or protect the plants from the worst of humid spells or drought, then their densely packed rosettes are something to see throughout the year, especially when there's no snow on the ground and their foliage outshines anything else in the garden.

Stonecrop Gardens has a section of their pit house dedicated to saxifrages in this group, but the greatest Kabschia collection I have seen lives at Waterperry Gardens in the United Kingdom. At Waterperry, these plants are given all they need to flourish at lower elevations. They're planted directly into Canadian tufa in raised beds, with armatures that support winter protection by keeping out the rain, followed by shade cloth to get them through the summer sun. It is an inspired plant collection.

In my version of a Himalayan garden, there are six weeks of intense bloom from mid-May to the first of July. While that may seem like an absurdly brief time to ask a garden to be eye-catching, it's not without interest the rest of the year. To me the garden is at its most beautiful in the cooler months, when the tight, undulating mounds of encrusted saxifrages sparkle. Each leaf on the dense basal rosette of *Saxifraga paniculata*, the lime-encrusted or White Mountain saxifrage, is edged with silvery white pores. When passing by in November or April, I find it impossible not to admire its presence. From small starts, the more vigorous forms mound up over rocks and spread across the pea stone. The silvery rosettes of *Androsace studiosorum* are a perfect complement. While the evergreen mats of *Gentiana acaulis* may brown out at times, these carpet-forming plants, unified by a bluish gray stone, have transformed a limited collection into a miniature alpine landscape.

Another early feature is drumstick primula (*Primula denticulata*), which bolts out of the ground with the melting snow. Its toothed leaves and farinose stems send up large globes of blue flowers, making it the showiest May-blooming perennial. A group fills the far end of the border, and would soon march along its length if their stray seedlings weren't picked out of the pea stone. There is some variation in the strain I have, with a few purple

CLOCKWISE FROM LEFT *Saxifraga paniculata* · *Androsace studiosorum, Saxifraga paniculata,* and *Gentiana acaulis* run the length of the bed. · *Primula farinosa* and *Gentiana acaulis* among tufts of early- and late-blooming saxifrages.

flowers among the blue, and these have been augmented by the closely allied *Primula capitata* var. *mooreana*. As the season progresses, their foliage enlarges, and with a little grooming to remove browning leaves, the cluster makes for a robust end point of the garden set off by the lady ferns that run its length.

The violet-blue trumpets of *Gentiana acaulis* succeed the blue drumstick primulas. These finicky bloomers sat quietly in the Rock Garden for years, every so often pushing out a flower or two. It was only when I moved them into full morning sun with moist soil that they began to flower profusely and form themselves into mats. There are a number of related species, and once I realized they had found a home, I tried a few selections obtained mostly from Edelweiss Perennials, a mail-order specialty nursery in Oregon. *Gentiana* 'Holzmann', 'Maxima' and 'Undulatifolia' have all done well, the latter appearing to be the most floriferous.

TOP LEFT Eastern Asian *Campanula chamissonis* spreads among *Androsace* and saxifrages.

BOTTOM LEFT From the Caucasus Mountains, *Gentiana septemfida* var. *lagodechiana* tolerates warm summers and is a reliable August bloomer.

MIDDLE Japanese *Gentiana scabra* var. *saxatilis* is the main flowering interest in this part of the garden in September. It has formed a sizeable clump, and seeds into the gravel.

FAR RIGHT In fall, the foliage of *Salix* ×*boydii* and *Gentiana scabra* var. *saxatilis* fades to russet and brown, while evergreen epimediums take on a tinge of frost and cushions of saxifrages tighten into a wintry dome.

Another primula with farina (a natural powdery coating of the leaves), diminutive *Primula farinosa* blooms a delicate pink next to the intense blue of the gentian, along with a few *P. marginata* blooming in soft lavender.

This late May blue phase wasn't intentional, but serendipity had it that the large *Syringa vulgaris* 'President Lincoln' lilac that presides over this part of the yard blooms blue with a tinge of pink at the same time, and this is reflected in the blue- and pink-blooming gentians and primulas. There is added adrenaline with the vivid pink and strongly fragrant *Daphne cneorum* 'Eximia'. Compact *Androsace villosa*, with its white flower and pink eye, makes a place for itself too. Finally, the lime-encrusted saxifrages get in on the accidental color scheme, covering themselves with flowers in white, rose, and pink.

While the vigorous flowering of the saxifrages is welcome, they earn their keep more in their foliage. What the flowers lack in subtlety they make up for

in exuberance. The relatively small flowers, held in a white forest of stems from 3 to 10 inches long, form a white haze the length of the bed. This bed is at its peak as the white mounds of saxifrages are joined by pink carpets of *Androsace* and the intense blue of trumpet gentians, all of it backed by lush green foliage of ferns and epimediums, and the fuzzy silver leaves of alpine willows.

A few hybrid *Dianthus* follow in mid-June, along with *Campanula chamissonis*, which continue the blue and pink color scheme. Taller *Corydalis elata* from the Himalayas shows a deep sky blue flower even on hot, muggy days of early July. Two late bloomers finish the primula season: the pink candelabra type *Primula bullesiana* and yellow *P. florindae*. In the cooler, shadier end of the garden, *Haberlea rhodopensis* and *Ramonda myconi* have found a place together.

By July, with some judicious deadheading of the saxifrages, primulas, and *Androsace*, the garden becomes tranquil again, showing off the gray foliage of

ROCK GARDEN
PLANTS
THAT SUCCEED

These genera of alpine plants have proven showy, hardy, and long-lived in the Rock Garden here:

Allium

Anthericum
(syn. *Paradisea*)

Campanula

Cyclamen

Epimedium

Gentiana

Phlox

Primula

Salix

Saxifraga

CLOCKWISE FROM TOP LEFT *Primula capitata* var. *mooreana* · *Allium thunbergii* 'Ozawa' · *Campanula cochlearifolia* · *Phlox subulata* seedling · *Saxifraga paniculata* · *Cyclamen purpurascens* · *Paradisea liliastrum* (syn. *Anthericum liliastrum*) · *Epimedium* 'Spine Tingler' · *Salix candida* 'Silver Fox' · *Gentiana asclepiadea*

Salix ×boydii and evergreen mats of daphne. There are occasional eruptions of pink and blue, including *Gentiana* 'True Blue', as well as a flush of pink from *Cyclamen purpurascens* with two other late-season gentians: *Gentiana septemfida* var. *lagodechiana* in August, and *Gentiana scabra* var. *saxatilis* that carries the garden through September. The bees are grateful for both of these late bloomers. After they're gone, the saxifrages sparkle on their own for a couple of months in anticipation of once again being covered by snow.

companion gardens

Surrounding the original Rock Garden are three closely allied gardens that make for a cohesive composition: the closest is a shade bed under an apple tree that is a continuation of the Rock Garden, with drifts of *Tiarella cordifolia* and collections of hellebores, epimediums, and ornamental ferns. Off to the side, on a bank above the Rock Garden, is a broad sweep of Russian cypress with paper birch and bottlebrush buckeye that conceals a small woodland behind it. The former stable, a step down from the milking parlor, is a green refuge, planted with swaths of alpine willows and native junipers.

UNDER THE APPLE

Two large apple trees at the roadside have limbs that stretch wide and high. In fall, they're covered with large, shiny green and yellow apples perfect for cider. In the areas closest to their sizeable trunks, the only plants that willingly colonize the dry, root-bound soil are *Tiarella cordifolia* on the shaded north end and *Asarum canadense* on the brighter southern side. A stand of interrupted ferns transplanted from a ditch by the roadside was an easy first effort to fill this space. Hellebores, epimediums, celandine poppies, rhubarbs, Japanese forest grass (*Hakonechloa macra*), *Actaea*, and *Leucosceptrum* have settled in where the root zone is less dense. These plants are well matched to the conditions, and all this planting needs to sustain itself is the occasional subtraction of some foam flower, ginger, or shrub mint. Because these plantings are fairly

Tiarella and ferns inhabit the drier soil under the apple tree, while the perimeter hosts epimediums, primulas, and other woodland plants.

stable, it has been a particularly good place to naturalize snowdrops. It's one of the first areas to lose snow in spring, and these little bulbs often bloom just as the snow recedes, followed quickly by a Tibetan hellebore, *Helleborus thibetanus*.

Along the perimeter of the apple bed, where they can be viewed up close, is a collection of ferns, epimediums, primulas, and choice woodland plants that include some spring ephemerals. A Himalayan rhubarb (*Rheum australe*) anchors one end of the bed, and contrasts with the delicate foliage of Himalayan maidenhair fern (*Adiantum venustum*).

Also along the edge grow *Primula juliae* and *P. sieboldii*, one in vivid pink and white and the other in purple; there is also a robust variety with a "hose-in-hose"-type flower given to me by George Schoellkopf, founder of Hollister House Garden in Connecticut. The interior of the bed was once awash in candelabra primulas, but the more rugged perennials have covered their territory, and now only a few of their brilliant red candelabra stalks appear along the margins in June and July. I gather seeds from the best, and plan to start them in a more propitious spot.

I've positioned a few tall, showy ferns in this area so they can be admired for their special habit. *Dryopteris crassirhizoma*, the thick-stemmed wood fern, is semi-evergreen with a strong upright presence. It rewards close inspection, with scaly croziers huddled at its base. Another wood fern, *D. affinis* 'Cristata The King', is the chief of the male ferns for its luxuriant green fronds with perfectly crested tips, all arranged symmetrically. It is easy to grow and readily shared with other gardeners, and has a companion here in perfectly formed maidenhair fern. Deep in the bed, a couple of false hellebore

FROM LEFT *Tiarella* and maidenhair fern, both native to the surrounding woods, on the slope shaded by the apple. · Hellebores, epimediums, rhubarbs, Japanese shrub mint, Japanese forest grass, Himalayan maidenhair fern, and *Polygonatum humile* have settled in where the root zone is less dense. · *Anemonella thalictroides* 'Snowflake' blooms in a quartet with white-flowered *Primula sieboldii*, *Tiarella*, and white-flowering epimediums. · Fresh foliage of *Leucosceptrum japonicum* 'Golden Angel', *Boehmeria platanifolia*, *Rheum palmatum* var. *tanguticum*, *Actaea simplex* 'White Pearl', and Japanese forest grass remains verdant through summer.

(*Veratrum viride*) unfurl early in the year, flower, and then disappear from attention. Other ferns that hold their own through summer include cultivars of maidenhair fern, *Adiantum pedatum* 'Billingsae' and *A.* 'Miss Sharples', along with the wood ferns *D. affinis* 'Crispa Gracilis' and *D. labordei* 'Golden Mist'.

After the rush of epimediums and primulas bloom, the garden under the apple goes quiet in late spring while the more robust summer woodlanders gather steam. *Deinanthe caerulea* and *D. bifida* send up hydrangea-like foliage, and later in July unfurl blue blossoms to match. *Anemonopsis macrophylla* sways above the shorter epimediums, rivaled by an exceptionally early–flowering *Tricyrtis*. A small votive sculpture of granite, wood, and lichen by sculptor Herb Ferris rests atop a granite slab that, in some summers, is also a setting for container-grown ferns or broadleaf foliage plants.

With adequate rainfall, cool nights, and some grooming, the garden under the shade of the apple is a restful carpet of green textures until fall, when *Leucosceptrum japonicum* 'Golden Angel', *Boehmeria platanifolia*, *Actaea simplex* 'White Pearl' (syn. *A. matsumurae* 'White Pearl') finish the season in September and October. 'White Pearl' provides pollinators that are still active in October a late-season supply of nectar.

A SECLUDED WOODLAND

Forming a backdrop to both the Rock Garden and Stable is a carpet of Russian cypress (*Microbiota decussata*) out of which grow paper birch (*Betula papyrifera*) and a mass of bottlebrush buckeye (*Aesculus parviflora*). This simple combination has pragmatic and design motivations behind it. The practical effect is that it screens the Rock Garden and Stable from the road; the aesthetic is that it's a foil for the plant-rich, colorful tapestry of the Rock Garden. The deep green of Russian cypress hugs the ground above the Rock Garden and Stable, giving rise to the white stems of the birches and midsummer white candles of bottlebrush buckeye, set off by its compound, coarse leaves. Both the cypress and buckeye break the edge of the wall, blurring hard architectural boundaries, softening them with foliage.

This combination does double duty in that the Russian cypress grows well in the open, sunny expanse, and makes for a feathered edge under the buckeyes and birches, but it is shaded out under them, leaving the interior open for a woodland planting. While the space under the buckeye isn't large, it's generous

LEFT A mass of bottlebrush buckeye rises out of a skirt of Russian cypress, forming an edge for the Rock Garden and concealing a woodland path beyond. MIDDLE Horizontal branches of bottlebrush buckeye create a shaded woodland setting for Christmas fern, sedge, epimedium, bloodroot, and trillium. RIGHT Mosses and Japanese forest grass aglow in cool autumn sunlight.

enough for an inviting path into a woodland scene, with naturalized trilliums and bloodroot, simulating a glade in the woods.

Trillium grandiflorum is easily divided when in bloom, and hard as it may be to imagine disturbing trilliums in flower, regular division has helped establish a trillium-rich woodland here. Bloodroot, on the other hand, spreads with no help from me. Growing deep under the birches are clumps of the sedge *Carex appalachica* and vigorous *Epimedium ×versicolor* 'Sulphureum'. Under an apple are a few additional sedges, along with evergreen Christmas ferns (*Polystichum acrostichoides*), some low-growing Solomon's seal and other single trilliums—*T. undulatum* and *T. luteum*. *Galax aphylla* hangs on at the edge of a rock under the apple. Come fall, as the days shorten and become moister, tufts of mosses glow and Japanese forest grass is burnished in the waning sunlight. This is a part of the garden whose care is mainly about editing, making sure to thin out the sedges and ferns that make themselves at home in beds of moss and leaves.

The most rewarding plants are those that can be arranged to grow in communities that appear natural and inevitable, and do not readily show the hand of the gardener.

Hakonechloa macra

Polystichum acrostichoides

Deinanthe caerulea

Primula kisoana

Primula sp.

Dryopteris crassirhizoma

Primula sieboldii purple form

Galanthus elwesii

Veratrum viride

Helleborus thibetanus

Epimedium sp.

Rheum australe

THE STABLE

As exciting and intensely colorful as the Flower and Rock Gardens can be, I am just as drawn to quiet, secluded spaces where I can be surrounded by green plants and lift my eyes to the sky overhead. The former stable is such a place: a favored destination with a bench to rest on, simple plantings of field juniper and alpine willow, lichen-encrusted granite stone walls, and glimpses of distant hillsides through a grove of birches and poplars. There is no place in the garden I feel more content, more at peace.

This hasn't been especially easy to achieve, and it's ironic that the one place in the garden that feels most natural is in fact a square room, bounded on three sides by granite and concrete foundation walls, and by a hedge on the fourth. Low, broad sweeps of field juniper, alpine willows, and cranberry play off the four walls. It is this simplicity, with plants in natural harmony and balance, that fosters a sense of repose and serenity. Birches and poplars that frame the view, and the apple trees and native shrubs on the other three sides enhance the feeling of seclusion.

LEFT The former stable offers seclusion and privacy, with *Salix nakamurana* var. *yezoalpina* as groundcover in the foreground. RIGHT Ironically, the place in the garden that feels most natural is a square room bounded by granite walls and a barberry hedge.

From the vantage point of the bench, the eye moves along the sweeping curves of junipers and willows, following mossy paths toward a break in the hedge, and outward through the arching branches of the trees to ridgelines on the far side of the valley. There is much for the eye to take in, and places for it to rest; no one feature draws attention to itself or is more important than the picture as a whole. It is a composition that doesn't tell a story, a landscape painting without narrative—it is a contemplative garden without any suggestion of religion.

One source of inspiration for this garden is the bald granite ledges on Cardigan Mountain, where low-growing conifers like juniper; trailing, ericaceous blueberries and cranberries; and a variety of grasses sprout from granite fissures. The fact that the foothills of the White Mountains can be glimpsed from the bench makes this inspiration feel more grounded in reality. It is a rather constricted list of plants that can withstand the short growing season and fierce winds of the mountaintops, and so it is in the Stable.

Ellen Shipman called privacy the "most essential attribute of any garden," and there is nowhere more secluded in my garden. The green space of the Stable is protected from the road and not readily seen from other parts of the garden, and it's the place I retreat to when I need to be alone with my thoughts. I am reminded of Charles Platt's remark on Italian gardens, that the garden and the grove were "where one wished to be at home while out-of-doors, where one might walk about and find a place suitable to the hour of the day and feeling of the moment, and still be in that sacred portion of the globe dedicated to one's self."

A formal Italian garden might be composed only out of stone, evergreen foliage, and sky, and this more naturalistic garden demands an equally rigorous approach. The near-absence of flowers, simplicity of the planting, and the calming effect of masses of green makes this feel more elemental and less like a garden. Stone, plants, and sky are what make this spot a garden.

Yet this has not come about by accident. It took a number of years for the plantings to mature, for that feeling of inevitably to take root, and for the intentions and materials of this area's design to become less evident. Some observation and adjustment along the way were needed to encourage the space to feel more authentic, less designed. The foundation plants of juniper, Russian cypress, cranberry, and *Salix nakamurana* var. *yezoalpina* have matured and knitted themselves into one another the way they might in a natural setting. Not all of these are native plants, but their habit and ability to combine with native plants makes for a credible sense of being in nature.

The granite foundation walls of the Stable are some of the best built on the property. They're covered with mosses and lichens; a few are cracked, suggesting exposure to fire at some point. These walls form the south and west sides of the Stable. The east side is a low concrete wall with a step up to the milking parlor and Rock Garden. The north side was open, with an earthen ramp leading into the Stable, but has now been closed off with a hedge.

The space's proportions suggested it could be developed as a garden room with a place to sit—at 28 feet square it's comparable in size to the oldest part of the house. I found a bench and set it against the stone walls in the southwest corner, shielded from the road under the spreading limbs of two wild apple trees.

The Stable and dairy foundations are neighbors, so these two areas needed to be both distinct and related to one another, each with its own interpretation of wild plantings. The Stable promised a garden experience very different from the tapestry of the Rock Garden. The Rock Garden would be colorful and horticulturally rich, while the Stable would be a departure from the excitement of color, and the work of tending flowers.

Sweeps of juniper, alpine willow, Russian cypress, and cranberry play off the walls. The simplicity of this planting fosters a sense of repose and serenity.

The bench's placement set a number of things in motion. There would need to be a step entering from the milking parlor, and stepping stones arranged to allude to a path to the bench. There would need to be a passageway through the hedge to the fields, and screening between the Stable and the road.

I moved a large granite slab into position to serve as a step between the Stable and Rock Garden, and laid stepping stones in a suggestive but not fully regular pattern. I picked through stones on the property and added 12 or 15 flat stones in a way that I hoped looked like remnants from some prior occupation, knowing that in time many of them would be obscured by moss and spreading groundcovers. The space was cleared of brambles and the grade leveled with the addition of sand; a well-draining, less fertile soil would be closer to the thin, bony soil found on the mountaintops.

Two apple trees, one on the southwest corner and one on the northwest, furnish screening with their foliage, craggy trunks, and thick, spreading branches. They serve to separate the space from the road, but more protection

Although the Stable is adjacent to the rock garden, I wanted it to be a green world, away from the rock garden's colorful tapestry.

was needed. For that I planted a number of staghorn sumac (*Rhus typhina*) in the rubble-filled rectangle between the Stable and road, counting on its tendency to sucker and fill empty waste spaces. These shrubs' compound leaves were not only good as a screen, but also contributed a native wildness to the scene. Unfortunately they were flattened in a wind shear and didn't recolonize, so I replaced them with bottlebrush buckeye. The buckeye looks less exotic, but its dense palmate foliage and white flowers provide an attractive screen. Growing out of it are paper birches to extend the birch and poplar grove along the roadside.

Behind the bench, spanning the two apple trees, I planted a run of *Hydrangea arborescens*, the native smooth hydrangea. It prospers in shade and its dense, coarse foliage is a visual stop when massed, enlivened by its dusky white lacecap flowers. A Korean mountain ash (*Sorbus alnifolia*) casts more shade behind the hydrangeas. Clothing the top of the wall is a mass of

FAR LEFT American cranberry grows in the dry soil of the Stable, threading through *Salix nakamurana* var. *yezoalpina*. Maidenhair fern and *Carex stricta* grow at the base of the wall. Russian cypress and bottlebrush buckeye lap the top.

MIDDLE A mossy path leads from the Rock Garden to the bench between cranberries and Russian cypress.

LEFT While yuccas and *Schizachyrium scoparium* 'Heavenly Blues' don't grow wild on northern New England mountaintops, they're at home in the Stable.

Hakonechloa macra, the seldom-used straight green species of Japanese forest grass, and for contrast, a clump of *Rodgersia podophylla*. The effect is to allow the eye to linger among the refined foliage textures along the top of the wall, and then to dissolve into the trees beyond.

For the interior of the Stable, the inspiration of ridge-top vegetation from the foothills of the White Mountains meant limiting myself to a short list of evergreens: *Juniperus communis* 'Effusa' and Russian cypress (*Microbiota decussata*); American cranberries (*Vaccinium macrocarpon*); and alpine willows (*Salix nakamurana* var. *yezoalpina*, *S. hylematica*, and *S. myrsinites*). These foundational plants take up the bulk of the sunny Stable floor. A few other woody plants find places around the periphery, including bayberry (*Myrica gale*), sea buckthorn (*Hippophae rhamnoides*), fragrant rhododendron (*Rhododendron arborescens*), and an oakleaf hydrangea cultivar (*Hydrangea quercifolia* 'Snowflake').

LEFT *Yucca filamentosa* RIGHT The moist corner by the bench is inhabited by the southeast native umbrella leaf, which has blue berries in July. OPPOSITE Small rosettes of *Saxifraga paniculata* were patched into crevices with mud and moss to keep the roots moist until they worked their way into the granite fissures.

The grasses on top of Cardigan Mountain would include low-growing fescues and sedges. Many fescues don't respond well to garden culture in more humid environs, so I decided to substitute other grasses that might better suit the Stable. These included mosquito grasses (*Bouteloua gracilis*) for their fescue-like bluish gray clumping foliage, as well as the tussock-forming sand lovegrass (*Eragrostis trichodes*). Little bluestem (*Schizachyrium scoparium* 'Heavenly Blues') turns redder as the season cools off, and *Sesleria autumnalis* clings to the base of a shaded wall. A few sedges rise out of the willows, the most dramatic of which is a tussock sedge, *Carex stricta*. Less prominent are two palm sedge cultivars, *Carex mushkingumensis* 'Oehme' and dwarf 'Little Midge'.

I attempted to grow mosses in the open sunny expanse between the junipers, and for a while succeeded with some native mosses. Repeated sprays of buttermilk and vinegar to acidify the soil helped get them established, along with an occasional application of glyphosate. The latter proved more effective at getting them going, a trick I observed at a conifer nursery in northern Vermont, where mosses thrived at the base of woody plants repeatedly sprayed with glyphosate. The open areas are now covered by a variety of mosses, and Irish moss is the latest addition to the mix, having escaped from nursery-grown plants—an acceptable disappointment.

Early on, I was almost seduced by the idea of a fountain or some sort of object as a focal point for this space. I soon ruled this out because anything showy would feel artificial, a striving for artistic effect that wasn't appropriate here. An agricultural remnant might be more acceptable, but it would have to be an honest piece, either found on the property or something humble enough that fit in. Some friends passed along a round, rusted tin pan, 5 inches deep and about 4½ feet wide. It sits comfortably nestled in the junipers with no specific purpose, and doesn't call undue attention to itself.

The junipers and rusted pan suggest a hot, sun-drenched clearing, and that impression is helped along by some of the secondary plants that have found a place. Common mullein is a welcome weed, and often seeds itself into spaces that could use some interest but might otherwise be difficult to plant. A little judicious editing in early summer reduces its dozen or so seedlings to a few well-positioned ones. Its soft gray leaves soothe the eye and combine well with the junipers. A lone yucca (*Yucca filamentosa*) and a small patch of

Illuminating the October garden, a gingko plays a leading role in the Rock Garden, Stable, and meadow.

OPPOSITE "That sacred portion of the globe" Charles Platt refers to, for me, is here.

eastern prickly pear (*Opuntia humifusa*), both U.S. natives but neither native to Vermont, are at home in a part of this garden that never receives supplemental water.

The shaded nook by the bench is cooler and moister, and is home to a few small plants that have all run together, making for an evergreen ensemble: *Epimedium platypetalum*, wintergreen, deer fern, and mosses. Sheltering this low-growing mix is the much taller umbrella leaf (*Diphylleia cymosa*), with spreading foliage at 4 feet. Its nondescript white flowers in late spring turn into blueberry-like clusters on cerise stems in midsummer. It has discovered exactly the conditions it prefers, and has seeded itself in to become a generous pillow by the bench.

It took a long time for me to take advantage of the growing conditions the north-facing granite walls here presented. The crevices between the stone offered a perfect spot for saxifrages. I pried out the lady ferns that dominated the wall and began patching in saxifrages among the moss-covered gaps between the stones, taking divisions from some of the older, more vigorous rosettes growing in the Rock Garden and troughs. First, I dabbed small clumps of heavy, moist loam into the sides of the crevices. Then I took a small bit of the same moist loam and patted it around the exposed roots of the saxifrages, pressing this root mass into the loam in the pockets. The interior loam held the root ball in place, and provided initial moisture and contact while the saxifrages grew new roots farther into the walls. I did this in spring and early summer, and made sure the roots remained moist through natural rainfall or occasional light watering during the plants' first summer. Most of the clumps have flourished and grown additional rosettes, which have worked their way along the crevices. The failures that have occurred have happened when the soil dries out during extended spells of heat and drought.

The bench here receives the most use during the cooler days of September and October when the evergreens deepen in color, as do the grasses. Outside the hedge, a ginkgo lights up brilliant yellow against the clear autumn sky, and the birches gradually shed their yellowed leaves. New England autumn paints the New Hampshire hillsides at the higher elevations first, then gradually moves downslope. Finally, by late October, the last of the leaves are gone, and the hillsides are etched by the gray and brown trunks of poplars, oaks, and maples, highlighted by white birches and dark greens of pines and hemlocks. The setting sun rakes across the hills and valleys, drawing them in starker contrast. There is a chill in the air, but I have found my sacred portion of the globe.

a living laboratory

As much as my garden encompasses the old fashioned, it is also a space to workshop contemporary styles—a laboratory to experiment with plants and fresh ideas.

Stepping out of the Stable through the hedge into the New Meadow. The dark foliage and right angles of the barberry hedge create an architectural frame for the looseness of the meadow planting.

Newer portions of the garden, where the New Meadow planting (center) transitions into an area dubbed Silver and Gold for the variegated dogwood just visible on the right.

The farther I've ventured from the house and Flower Garden, the more adventurous my gardening has become.

New garden areas inspired by the bold gardeners of the Pacific Northwest provide settings for plants that might be jarring in close proximity to the garden's core, but offer surprise and delight away from it. I call one such garden "Silver and Gold," after its namesake, a cultivar of the dogwood *Cornus sericea*. Like that plant's foliage, it features variegated silver, white, and yellow shrubs and perennials, as well as trees and shrubs selected for winter interest.

Another planting, the New Meadow, takes its cue from the gardens of Dutch garden designer Piet Oudolf and his contemporaries, loosely associated with the New Perennial Movement. It is composed of a meadowy mix of grasses and persistent perennials. Closer to the heart of the garden, but in high contrast to the colorful summer Flower Garden, is a sumptuous border of hardy plants chosen for their bold foliage, simply named the Long Border.

silver and gold

It's not lost on me that the most innovative part of my garden, and a fairly extensive one at that, wasn't part of my original concept for the garden or its master plan. Silver and Gold took shape only gradually, with no specific goal or direction at the beginning. But it has taken on its own personality, driven at first by my desire to collect plants and try to find suitable places for them, so that eventually a theme emerged.

The planting began as a backdrop for the Stable, with a grouping of smooth *Hydrangea arborescens*, intended to confer a sense of privacy and enclosure. This plant is the native species that has given rise to 'Annabelle' and other new horticultural varieties. I like the denseness of its broad leaves and its delicate lacecap blooms, and find it useful en masse as an endpoint to a garden. I first planted a group on a mound of what had been a compost pile on the Stable's far side. I also took advantage of this mound to place a few other plants that I wanted to gain in size before transplanting elsewhere, positioning them on the far side of the hydrangeas, out of view of the Stable.

A few of these plants did well and actually made sense where they were first placed. They included willows, goldenrods, and a perennial sunflower, *Helianthus* 'Lemon Queen'. Nowhere else in the garden did I have such yellows, and they looked great at its wilder end. A couple of *Amsonia* and an ornamental rhubarb were added, and all of a sudden a perfectly fine garden began to evolve out of plants that weren't working elsewhere.

The view back to the hills and valley pulled me out here, but once the field grasses had grown high, there was little reason to venture to this part of the garden. And so the idea for a new planting took shape, one that would draw visitors up the slight slope and reward them with an inspiring view back over the garden.

There is timelessness in this view: in the foreground, there is a wild apple tree, and beyond that the grass path and Lombardy poplars in the distance that draw the eye across the valley, toward the long ridge and expanse of sky. Here is one of the best vantage points for observing the gently sloping ridges of the river valley, echoed in open fields and wooded hillsides. Because the land gradually falls away, the view of the sky is even more dramatic. The long ridge of Moose Mountain frames it as if it were a painting.

I planted this garden as if I were composing a landscape painting. With the poplars in place, another columnar plant was needed in the foreground. I chose a narrow, upright cedar, *Thuja occidentalis* 'Rosenthallii'. I placed a mugo pine at its foot soon after, and then finished the evergreen framework

LEFT *Thuja occidentalis* 'Rosenthalii' repeats the upright forms of cedars in the main part of the garden and contrasts with the silvery tones of willows *Salix candida* 'Silver Fox' and *S. alba* var. *sericea*. **MIDDLE** *Cornus sericea* 'Silver and Gold' introduces this area's theme with its variegated green and white foliage and yellow winter stems, rubbing shoulders with *Hydrangea arborescens* 'White Dome'. **RIGHT** *Cornus sericea* 'Silver and Gold', underplanted with *Picea abies* 'Repens' in the foreground and *Hakonechloa macra* 'All Gold'.

with a spreading Norway spruce (*Picea abies* 'Repens') at the farthest end. The grass was still tall when these were planted, but I could tell as soon as they went in that a garden could compose well in this location.

Next came a number of shrubs I had collected that were growing in the nursery and ready for placement. These included a silver-leaved willow, *Salix alba* var. *sericea*, with narrow metallic foliage, which was planted next to another dark green mugo pine for contrast. Then came *Cornus sericea* 'Silver and Gold' for its strikingly variegated leaves and yellow stems in winter. Just beyond the variegated dogwood, the gold foliage of *Robinia pseudoacacia* 'Frisia' is a beacon, and silver-leaved hawkweed (*Hieracium pilosella*) ties them all together as a groundcover.

'Silver and Gold' is one of many plants in my garden introduced to cultivation by Dr. Richard Lighty. A friend and mentor, Dick was the first director

of the Mt. Cuba Center in Delaware, where he chanced across a sport of variegated foliage on *C. sericea* 'Flaviramea', which he carefully cultivated and introduced as 'Silver and Gold'. Among many other plants Dick brought to cultivation are *Hydrangea serrata* 'Blue Billow' and *Aruncus aethusifolius*, from which he collected seed in South Korea in 1966. These plants have pride of place here, both for their superior garden worthiness and because of my friendship with Dick.

It gradually became apparent that a theme was emerging—a composition of foliage in silver, gold, and harmonious tones. It's no accident that this garden became a little more adventurous soon after my work took me to other parts of the country, especially the Pacific Northwest. Plant combinations there were more exuberant and less constrained by traditional notions. The region's cloudy skies are also more conducive to imaginative uses of foliage—and I find gardeners there more adventurous in temperament.

The garden that opened my eyes more than any other to the possibilities of working with colorful and variegated foliage was Bella Madrona, a private garden in Sherwood, Oregon, created by Geof Beasley with Michael Schultz. To call it vivid does not do it justice—what I saw there was a fearless intelligence at work combining plants for effect based on their foliage. Conifers, shrubs, and perennials worked together to produce many a memorable scene and arouse emotion with explosive color. The range of plants available to gardeners who favor foliage in the Pacific Northwest is mind-boggling, and many practitioners there do it well, but this is where I first saw such vivid color used in a living garden and not as a stage set. It brought to life the possibilities in my own garden. I discovered there were many plants offered for sale in the Northwest that performed perfectly well in New England.

In Silver and Gold, two large *Cornus* 'Silver and Gold' billow forth under the spreading limbs of an apple tree at the entry to the border, their lower limbs swept by gold Japanese forest grass (*Hakonechloa macra* 'All Gold'). This grass also helps tie the border together; there is a drift of the green form with rodgersias as a transition between the New Meadow planting and Silver and Gold. Farther along, there's a spot of the white-leaved form (*H. macra* 'Albo Striata'), and a large clump of variegated *H. macra* 'Aureola' just beyond.

Hydrangea arborescens 'White Dome' is a year-round presence here. It's a strong grower with coarse foliage and, from July onward, covered with white inflorescences that are a magnet for pollinators. It grows to 6 feet or more, and its strong stems hold the plant's dried flower heads even through heavy snows. It also contributes to the border's surprising foliage display by allowing the yellow-leaved *Aralia cordata* 'Sun King' to grow through its branches, giving the group a two-toned effect. This was unplanned, but it's a happy result of my dense planting style (some might call it overplanting).

Off to the side is a clump of *Molinia caerulea* ssp. *arundinacea* 'Transparent', which allows for a view into the dense interior. A large stand of *Persicaria polymorpha* grows on one side, and rhubarbs and silvery willows on the other. A second view on the other side of the 'Rosenthalii' arborvitae features some of these same plants from a more revealing angle, and the addition of both a small, silver-leaved willow (*Salix candida* 'Silver Fox') and a yellow-leaved elderberry (*Sambucus racemosa* 'Sutherland Gold').

The garden opens to an alcove with a small lawn, where the tableau of Silver and Gold is on view in full frontal profusion: *Cornus* 'Silver and Gold' again, this time with the long stems of *Sambucus nigra* 'Laciniata' winding their way up through its variegated foliage, tricking the viewer into thinking the dogwood has large flat cymes of white flowers growing from its white leaves. Here the dogwood is backed by another taller, yellow-leaved dogwood

A stand of *Persicaria polymorpha* is held in check by surrounding shrubs and an ancient apple tree.

(*Cornus sericea* 'Hedgerows Gold') on one side and bottlebrush buckeye on the other. *Ptelea trifoliata* 'Aurea' rises farther back, with a variegated elderberry, red-stemmed dogwoods, and various willows, the last of which are mainly showy in winter months. This dense planting also includes upright junipers and arborvitaes to help screen it from the road. The foreground is a medley of silver-leaved artemisias, grasses, and variegated sedges for fine texture, as well as Jerusalem sage, hellebore, *Aralia cordata* 'Sun King', and *Symphytum* 'Axminster Gold' for bold, sun-drenched textures. A carpet of acid-yellow moneywort (*Lysimachia nummularia* 'Aurea') holds it all together. Some years, the tall, silvery biennial Scotch thistle (*Onopordum acanthium*) is allowed to enter the fray, but the showstopper is always a tall mullein from Great Dixter in England, *Verbascum* 'Christo's Yellow Lightning'. There are times when the yellows (especially of the moneywort) are harsh on the eyes, but the silvers of the willows and hawkweed are always soothing. Dark foliage of hellebores helps ground the composition.

I have used *Robinia pseudoacacia* 'Frisia', a black locust cultivar with lemon yellow foliage, both as a cutback shrub and a tree—one that was cut back for a few years is now being allowed to grow, putting on as much as 6 to 8 feet in a year. Its bright yellow leaves, which retain their color throughout the

LEFT In an alcove in Silver and Gold, the foliage of dogwoods, silvery willows, golden elderberries, aralia, and variegated comfrey intermingles with grasses, mulleins, and self-sown Scotch thistles. MIDDLE Silvery Scotch thistle, yellow aralia, *Hakonechloa macra* 'Aureola', and *Symphytum* 'Axminster Gold' rise above yellow moneywort, with the foliage and flowers of hybrid hellebores offering contrast. RIGHT *Verbascum* 'Christo's Yellow Lightning'

season, are a beacon from far reaches of the garden, and carry the theme from the ground level up the length of its 40 feet. Some years it is draped with white, acacia-like blooms, a subtle complement to emerging yellow foliage.

Long before the theme for this area became Silver and Gold, when all I was concerned with was creating a backdrop for the Stable, I planted a Korean mountain ash (*Sorbus alnifolia*) largely on the advice of horticulturists at the Arnold Arboretum in Boston. It is a tree that thrives in cool northern locations, and one of the few that is showy in fall for its combination of colorful foliage and striking red berries. That fall foliage can take on a very acceptable golden hue, so it's found its place within the theme. Mountain ash's spring bloom is reliable and robust, covering the tree in white flowers at the end of May.

Not everything in this part of the garden is choice and unusual. Some plants are unusual but outdated, like the adjoining false spirea (*Sorbaria sorbifolia*). It's a plant found most often around old farmsteads, super hardy and growing thickly—and aggressively. Its drawback is that it wants to run

RIGHT Korean mountain ash is covered with creamy white flowers in May, which mature into red fruits that remain on the tree through winter. The acid yellow of *Lysimachia* 'Aurea' is at its most jarring in spring.

MIDDLE Seen from the road, a mélange of aggressive woody shrubs creates a privacy screen and also carries the Silver and Gold theme to the far end of the garden. *Sorbaria sorbifolia* runs into *Elaeagnus* 'Quicksilver', along with *Salix gracilistyla*, which merges into native thimbleberry.

FAR RIGHT *Thuja* occidentalis 'Malonyana' and *Robinia pseudoacacia* 'Frisia' seen from the road.

and become part of the border, yet I've kept it because it makes such a good screen, its dense suckering habit an effective barrier to the road year-round. Its spirea-like plumes in July add some midsummer interest, but in my opinion its compound, deep green foliage also makes it worth keeping.

The false spirea here shows off best from the roadside, so what I think of as the backside of the garden is also part of a deliberate arrangement viewed from the road. The plant runs together with another equally aggressive shrub, native thimbleberry (*Rubus odoratus*), which intermingles with red-stemmed dogwood (*Cornus sericea*), also native. Rising up behind these edge inhabitants are more silver- and gold-foliaged plants: rambling *Salix gracilistyla*, with its additional contribution of silver-gray catkins; the rambunctious but controllable *Elaeagnus* 'Quicksilver'; and a golden *Catalpa bignonioides* 'Aurea', treated as a cutback shrub. Two columnar evergreens, *Thuja occidentalis* 'Malonyana' and *Juniperus virginiana* 'Emerald Sentinel', add structure. Arching over them, the 'Frisia' black locust entices passersby from the road to enter.

Much of the early spring activity in the overall garden occurs at this end—it's where we used to stack downed limbs and woody trimmings in anticipation of a bonfire in early April before the snow is gone—so some winter interest here was called for. Upright evergreens stand sentry against the snow, enlivened by scarlet, yellow, buff gray, and brown stems of shrubby willows and dogwoods.

Yellow-stemmed dogwood (*Cornus sericea* 'Bud's Yellow') echoes the winter-yellow stems of *C. sericea* Silver and Gold, and a red-stemmed dogwood (*C. sericea* 'Alleman's Compact') offers some contrast. I tracked down and added a blue-stemmed willow, *Salix irrorata*, I first noticed in the winter garden at the Washington Park Arboretum in Seattle, as well as white willow varieties with gold and red stems, *S. alba* var. *vitellina* and *S. alba* 'Britzensis', respectively. All these are treated as cutback shrubs, although I sometimes allow them to grow to 25 feet. Their colorful stems and emerging catkins outline a tracery against the azure April sky.

LEFT Small gray catkins of the blue stem willow *Salix irrorata* dart around the yellow stems of *Salix alba* var. *vitellina* against a late-winter blue sky. RIGHT Tawny seed heads of *Hydrangea arborescens* 'White Dome' weave through yellow stems of *Cornus sericea* 'Silver and Gold', while late-autumn foliage of Japanese forest grass and a low-growing evergreen spruce hold the ground. OPPOSITE Fragrant, pure white flowers of *Narcissus* 'Thalia' among the emergent foliage of variegated dogwoods, Japanese forest grass, and hostas.

Dark green of juniper, arborvitae, and mugo pine alternates and contrasts with yellow, red, blue, and brown stems of the shrubs. The sight of the Korean mountain ash covered with red fruit against a crisp, blue February sky is breathtaking. Less striking but fully engaging are the dried, tawny seed heads of *Hydrangea arborescens* 'White Dome', especially when they sport a dusting of snow.

As the snow recedes, narcissus and squill push through the mud beneath the shrubs. I learned from Fergus Garrett, renowned head gardener at Great Dixter, how to create a display of spring bulbs by planting in groups under and around the base of permanent shrubs. Most of the dogwoods and willows shelter an ever-increasing parade of bulbs, timed to begin just as the snow melts and continue until the shrubs have fully leafed out. Narcissus cultivars are

some of the most successful here, including 'Sailboat', 'Jetfire', 'Hawera', 'Sun Disc', 'Topolino', 'Thalia', and 'Tête à Tête'.

Hybrid hellebores spread throughout Silver and Gold as underplanting. Early experiments with hellebores I brought back from the West Coast proved viable, and those that bloomed in purple, violet, and slate went best with this color scheme.

In Vermont, hybrid hellebores usually emerge from the snow in tatters, and take some time to make presentable with a little help from pruning shears as well as warming days and moderating nights. They make their best showing in May alongside many other, more traditional New England spring flowers.

Once I realized hellebores could be a reliable contributor to the garden, I continued with *Helleborus foetidus* 'Wester Flisk' for its deeply divided evergreen foliage. Although it has a reputation for being short-lived, it makes

up for this by seeding itself around in a respectable way. The Winter Jewels breeding program from venerable Oregon horticulturists Ernie and Marietta O'Byrne has resulted in most attractive hybrids in the violet spectrum, among them *H.* 'Pippa's Purple' and *H.* 'Amethyst Glow'. More variety is on its way, as these dark-flowered hybrids create new seedlings that place themselves in and among the spreading shrubs.

I grow a few hellebores for the rich coloration of the veining on their dark foliage. *Helleborus ×nigersmithii* 'Walhelivor' (also known as 'Ivory Prince') is a strong grower known for its sturdy, outward-facing, pink-infused buds and white flowers. The light veining on its dark green foliage welcomes it into this composition, as does that of equally floriferous *H. ×ballardiae* 'HGC Pink Frost'. The award for nicest veining in the group, however, goes to *H.* 'Winter Moonbeam'.

October light shines through the branches of an apple tree, illuminating the foliage of a recently planted Korean maple.

the cove

What was once an awkward dead end, with a few scattered plants seen only by those on their way to the burn pile, is in the process of becoming a part of the garden proper. The burn pile has been relocated, opening up a glade in the shade of an impressive apple tree.

Along the edge of the road, a red oak and a wild apple tree are the backbone of a mixed planting of woody shrubs, herbaceous perennials, and bulbs that is beginning to come into focus. An early spring-blooming currant (*Ribes odoratum*), clothes the air with its clove scent, followed by the true blue flowers of *Camassia leichtlinii* 'Caerulea', which rise up the stem and seem to be reaching for a white-flowered pagoda dogwood (*Cornus alternifolia*) blooming above them. A Korean maple (*Acer pseudosieboldianum*) is in the process of becoming established in this shaded nook.

The camassias are happy in the dappled shade of the dogwood and apple, and are increasing into mature clumps. These are intermingled among clumps of New England and New York asters, and a fall-blooming monkshood. After the camassias finish, the border goes quiet for summer, although we're drawn there to pick black currants that ripen in July. A late May or early June shearing of the asters helps keep them from becoming leggy.

For a long time, the asters and camassias were just unassuming stragglers, but now the planting has become a little more intentional. It perks up again in late September, when the asters begin their show. They're well suited to the wildness of this end of the garden, and also enjoy some shade from the afternoon sun under the apple and dogwood. A few of these asters have been with me for 25 years; others I've snagged from neighboring gardens. Most came without names, save for a few older stalwarts such as *Symphyotrichum novae-angliae* 'September Ruby', 'Woods Pink', and 'Andenken an Alma Potschke'. As cheerful as this corner is when the asters are in bloom, it's at its most elegant in October, when deep purple flowers of *Aconitum carmichaelii* lean alongside the buffed orange and red foliage of the Korean maple. There is no better combination of foliage and flower anywhere in the autumn garden.

The burn pile has been leveled and will be planted. A new path has been opened up so garden strollers can continue past the apple and to the orchard beyond. A bank of the impressive *Hydrangea arborescens* 'Haas Halo' has been installed to link the asters and camassias with Norway spruce on the property's edge. A visitor told me I should give this area a name.

I am approaching the design of the Cove slowly and deliberately: the first season was spent suppressing weeds and monitoring available

LEFT A selection of New England and New York asters (*Symphyotrichum novae-angliae* and *S. novi-belgii*) at the edge of the Cove. RIGHT A winning fall combination under the apple tree: Korean maple and *Aconitum carmichaelii*.

sunlight—discovering how the sun bakes it at midday—but the challenge has been getting a sense of what the space wants to feel like, and what plants would best contribute to that feeling. It feels like a cove sheltered by hills with a bright, sunny glade at its core. We shall see how it evolves.

Skirting the edge of the field, the New Meadow planting backed by rosemary willow and ginkgo connects with Silver and Gold. **OPPOSITE** The New Meadow is an experiment with a matrix of ornamental grasses and perennials chosen for their sociability. Shown here: *Salvia* 'Caradonna', *Stachys officinalis* 'Hummelo', *Eryngium* 'Blue Glitter', and *Achillea* 'Sunny Seduction', plus buds of *Allium* 'Millenium' and *Deschampsia* 'Goldtau'.

the new meadow

One of the overall garden's challenges is that any new addition must be thought through in order to assess how it will dovetail with the scale and composition of the garden and landscape as a whole. It's best to avoid making statements that are so small as to be insignificant, too large to maintain, or sited so as to detract from the appreciation of the greater landscape. For a while, I thought I had reached the maximum of gardened space on this site, and that there were no more appropriate new spaces to build out. Yet the urge to experiment with new plants and attempt newer garden styles is ever-present.

Although much of the garden is outfitted with old-fashioned plants and garden styles not currently in fashion, I don't allow that to define me as a gardener, but finding places to experiment with new plants and new ideas has been a challenge. For a number of years, I collected ornamental grasses and new perennials used in meadow-inspired plantings by Dutch garden designer Piet Oudolf and others, and tried to integrate them into the existing gardens. Frequent visits to the High Line in New York and a week spent touring gardens in the Netherlands and Germany drew my interest to the plants of the steppes and shortgrass prairies that this contemporary garden style makes use of. I struggled to find a place where I could introduce a garden inspired by these plants—it didn't seem quite right to create such a garden right on the edge of the existing meadows and hayfield.

Ultimately I found a place that did. Part of the High Line's success is that its loose plantings in all their variety play off the solid structure of steel walls, concrete pathways, and the surrounding architecture. I came to see that the experience of stepping out of the Stable onto lawn and the hayfield just beyond could instead be the experience of moving through a shortgrass meadow in the foreground, with lawn in the middle ground, and the field in the background. The space was large enough that a meadow-like planting could make sense, but not be so large as to require a whole new level of maintenance. I also saw that I wouldn't need to purchase many plants, because I could move grasses and perennials better suited to this type of garden out of less advantageous positions elsewhere. Yet I could also justify acquiring new and unfamiliar plants to learn how to work with them.

LEFT An Oudolf-inspired combination of *Salvia* 'Caradonna', *Monarda bradburiana*, and *Amsonia hubrichtii*, with *Verbascum phoeniceum*, *Amsonia* 'Blue Ice', and seed heads of *Pulsatilla vulgaris*. RIGHT *Amsonia* 'Blue Ice' features blue flowers in spring and lustrous green foliage in summer, turning golden yellow in fall. OPPOSITE *Pulsatilla vulgaris* is first out of the gate to bloom, and has the bonus of persistent, showy seed heads when warm-season perennials and grasses begin their season.

The site's one drawback is that it receives less than full sun. A mature apple tree on the Stable's northwestern corner and a 14-foot *Ginkgo biloba* I'd grown from a small sapling cast a fair amount of shade. But these drawbacks were also advantages: the hedge defining the north side of the Stable connected the ginkgo and apple, and created an almost architectural backdrop for the New Meadow.

Sue and I began by stripping the sod off the new bed and laying sheets of newspaper underneath the ginkgo, then covering it with bark mulch and shredded leaves to suppress the grass under the tree. A bed of barren strawberry (*Waldsteinia ternata*) had been planted under the apple years before, and we stripped the sod back to that. I didn't improve the soil at all, but simply

removed a few rocks and held back from planting long enough to make sure there were no weeds to emerge where the grass had grown. This work was done in autumn, in anticipation of planting the following spring.

Studying plantings at the High Line and reading about similar planting styles encouraged me to limit the matrix of grasses to only two or three primary varieties, and to plant them in the sunnier portion of the garden. I chose *Sporobolus heterolepis*, commonly called prairie dropseed, as well as *Deschampsia cespitosa* 'Goldtau' and *Sesleria autumnalis*. Sedges were the better choice for the shady end, where I used the native *Carex plantaginea* and Japanese *Carex* 'Ice Dance'. I've also allowed the barren strawberry to move through the sedges and perennials to cover the ground.

Perennials were chosen in part based on their "sociability"—their ability to coexist in a densely planted environment with a long season of bloom and combined flowering effects. Additional selections were included to attract and provide nectar to butterflies, moths, and a wide variety of bees and hoverflies.

The earliest colors in the meadow are the purples and reds of pasque flower (*Pulsatilla vulgaris*), followed shortly by towering spikes of a purple mullein (*Verbascum phoeniceum*), which is especially attractive to hoverflies. Although it is a short-lived perennial, this mullein seeds itself around, and its broad

LEFT Delicate blossoms of Bowman's root relish the dappled shade of the ginkgo and apple tree overhead. MIDDLE Magenta-pink flowers of *Geranium macrorrhizum* 'Czakor' in lively conversation with the apricot of long-flowering *Geum* 'Totally Tangerine'. The mellow white bloom of *Rodgersia pinnata* doesn't contribute much to the discussion. RIGHT *Achillea* 'Sunny Seduction' and *Allium* 'Millenium' hold either side of the path, while *Anthericum liliago* spills over it.

foliage lies flat against the ground, helping give coverage to the soil and retain moisture. During the weeks of mullein bloom, the pasque flowers mature into showy silky seed heads that dance above their silvery foliage.

Early summer display highlights three signature Oudolf plants: *Monarda bradburiana*, *Salvia* 'Caradonna', and *Amsonia* 'Blue Ice'. They go well either paired (especially the latter two), or seen at a distance, with their flowers of pink, purple, and blue spreading and rising among the grasses.

A group of Jerusalem sage (*Phlomis russelliana*) sends up tall tiers of whorled yellow flowers above evergreen mats of foliage. One of the goals of the meadow planting was to utilize drought-tolerant plants that would never require watering. This plant grows along the front of the border, by itself, in full sun. As its soft yellow flowers fade to papery brown whorls, other more colorful plants draw attention, but I leave the tiers of the sage alone to develop as they would in the dry, rocky fields of their native habitat.

Meanwhile, *Amsonia tabernaemontana* 'Salicifolia' and the New England native Bowman's root (*Gillenia trifoliata*) skirt the entry from the Stable under the shade of the trees. The delicate white flowers of the Bowman's root, like Matisse cutouts, remain in bloom for weeks. They fade once the heat of summer takes hold, but the plant's foliage stays fresh in shade and can go bronze in fall. The narrow, willow-like leaves of *A.* 'Salicifolia' are also reliably green throughout the season. Each plant is a welcome sight when emerging from the Stable to the meadow. Off to the side is a mass of *Geranium macrorrhizum* 'Czakor', which holds one end of the bed, its magenta flowers enlivened by the serendipitous addition of *Geum* 'Totally Tangerine'. The orange and magenta of this pair in early June portend the hotter colors of summer. Also contributing to the foliage show in this portion of the garden are *Rodgersia sambucifolia* and *Carex* 'Ice Dance'. Geranium and rodgersia supply heft, while *Amsonia*, Bowman's root, and sedge add refinement.

The upright, rosy lavender stems of *Stachys officinalis* 'Hummelo' continue the vertical accents of *Salvia* 'Caradonna', rising above scalloped green clumps of foliage. It is useful in this meadow as well as in the more composed Flower Garden. *Allium* 'Millenium' soon unfurls rose-purple globes above glossy green foliage. All manner of bees, butterflies, and moths are attracted to its long-lasting flowers; when they've faded in late summer and fall, they carry

FROM LEFT Strongly upright flower stems of prairie dropseed, along with *Deschampsia cespitosa* 'Goldtau', catch the autumn light against saturated tones of *Sanguisorba officinalis* 'Red Thunder' and *Symphyotrichum laeve* 'Bluebird'. • Ruby red thimbles of *Sanguisorba* 'Red Thunder' play off wheaten stems of *Deschampsia* 'Goldtau'. • A feast for tiger swallowtails: *Echinacea* 'Evening Glow' backed by *Allium* 'Millenium'. • Syrphid flies and bees are drawn to *Allium* 'Millenium'.

on with a purplish blush and softening tan. This allium grows robustly for me, and I don't cut it back until spring.

Eryngium 'Blue Glitter' and *Achillea* 'Sunny Seduction' are predictable and useful partners in the meadow, with the sea holly's metallic blue complementing the lemony yellow of the yarrow. Less predictable but even more welcome are the tall, delicate white flowers of *Anthericum liliago*. I labored to find its proper place in the garden until I introduced it into the meadow planting. Arching over the pathway, it invites a gentle brush away from passersby.

I avoided growing coneflowers in the Flower Garden's dense conditions but have tried *Echinacea* 'Evan Saul' and 'Evening Glow' in this area, and I'm still not won over. It may be too crowded for them here as well, and I sense they're not exactly at home in the partial shade and heavy soil.

The garden comes into its own when the grasses combine with late-season perennials. Two of the standout grasses are repeated throughout. Prairie dropseed is somewhat slow to establish but worth the wait. Its thin, wiry blades sway in the breeze and meet the ground as a skirt would; its soft plumey inflorescences invite touch. Adding grace notes are the flower spikes of a cultivar of a northeast native grass, *Deschampsia* 'Goldtau'. Its evergreen tussocks mound up, while its flower heads glow when backlit by the sun, offering many months of gentle movement. In some years, the flower spikes remain erect through winter, etching a faint trace on the snow.

Burnets send up their ruby red pincushions among the grasses. *Sanguisorba officinalis* 'Red Thunder' is another plant selected and rightfully much used by Piet Oudolf. It performs well enough in the Flower Garden, but is at its best in these meadow-like conditions as a companion to grasses and later-blooming perennials.

Toward the middle of the meadow, *Amsonia hubrichtii* is valued for its fine, grass-like foliage that combines with prairie dropseed, and although its small blue flowers attract some interest, it finds its place because its soft foliage turns a golden-straw color in autumn.

The flowering season mellows into its final show with *Symphyotrichum laeve* 'Bluebird' and *S. oblongifolium* 'October Skies'—but nothing equals the October gold foliage of the ginkgo, which gradually subsumes the tree and then flutters to the ground, where it fades with the rest of the planting, left untouched for winter.

This experiment has its successes and its failures. It has plenty of interest: many of its perennials have a long blooming season, and their fading foliage has an afterlife that plays well against the dancing seed heads of the grasses in fall. Moths, butterflies, bees, and hoverflies find nourishment as long as they are active. It would be better if the location was more generous in size and wasn't compromised by shade—a few of the perennials don't get the full sun they need to remain upright. And in a meadow, one wants to see patterns and repetition of form and color, and there is barely enough space to accomplish that in this planting. One disappointment is that heavy wet snows, which now arrive earlier and earlier, flatten all but the most stalwart stalks. Jerusalem sage and bee balm may remain upright, but most other plants, including the grasses, tend to be flattened and seldom revive. Nevertheless, there are times when my breath is taken away by some new play of light, color, or motion, even in the snow, and I'm grateful for this addition.

TOP A December medley of *Amsonia*, *Deschampsia*, coneflower, allium, lovegrass, and evergreen Jerusalem sage. BOTTOM Coneflowers, alliums, *Sanguisorba*, and *Deschampsia* retain some of their charm even after heavy wet snows.

the long border

One of the benefits of master planning is that it leads you to an overall concept and definition of space, and if it works well, it leaves room for adaptation and change. This is how it's worked out with the largest open and level space in the garden, an area known as the Long Border.

It is an open space, 75 feet long by 50 feet wide, with a 10-foot-wide border running its length, entered from the Flower Garden on the east side, bordered by a hedge paralleling the field's edge on the north, and a long, stone foundation wall from the former hay barn on the south. The westerly far end opens onto the field and orchard, and to the Silver and Gold garden. It was here I made my first attempt at a border with the excess phlox from the Flower Garden. That didn't last long—a winter with poor snow coverage severely set back the phloxes, and that became my excuse for ripping it out. It was a mistake to plant such a colorful border so close to the Flower Garden.

LEFT The Flower Garden gives way to a hedged enclosure with open lawn and 70-foot-long foliage border. RIGHT Restrained textures and greens of the Long Border are a pause between the Flower Garden and the assertive Silver and Gold beyond.

I began thinking about this garden in terms of the experience I sought to create. After the failed attempt at a phlox border, I knew I wanted to pass through the arch into a dramatically different space with its own powerful set of emotions. What eventually emerged was a border emphasizing foliage and textures of large-leaved perennials. I wanted it to be less reliant on the fleeting excitement of flowers and more lasting, with impact from bold, verdant foliage. I think of it as a tropical garden fashioned out of hardy perennials. It was planted when interest in tropical plants was at its peak, but bananas and cannas don't work well in a Vermont garden, and I thought I would have fun imitating the style with exotic hardy plants.

It is also a large, open place to engage with the sky and views beyond. A viburnum hedge emphasizes the view to the field and the long wooded horizon to the north. The Lombardy poplars and arborvitaes draw the eye up to the sky to the east. The stone foundation wall, planted with vines and a border of green foliage, makes for a calm backdrop under the shade of birches and

poplars. Although the master plan set up the space, I didn't foresee the mood it could create.

I'd made a start with a planting of *Stephanandra incisa* 'Crispa' at the far end of the stone wall to obscure the hard edges of some large retaining stones. At the end closest to the Flower Garden, I planted a mass of *Rheum australe*, the largest, boldest-leaved rhubarb I knew. This plant was passed on to me by Massachusetts gardener Sheila Magullion, left over from a flat of seedlings she'd grown as a volunteer at the Arnold Arboretum with seeds sent from Beijing Botanical Garden. The ground roots of *Rheum australe* were a major export crop in China in the 19th century for treatment of digestive, respiratory, and circulatory ailments. Its leaves grow 3 to 4 feet wide, and the plant can reach 7 to 8 feet high when its flower stalks rise up. Four large plants have fanned out against the wall here for 20 years.

Slowly the border took shape. A little beyond the rhubarbs, I planted *Euphorbia palustris* 'Walenburg's Glorie', a plant I'd found at Heronswood in

LEFT A place to pause, where views surrounding the border are every bit as important as the rodgersias, rhubarbs, and *Euphorbia* here. MIDDLE A tropical garden made of hardy shrubs and perennials: *Magnolia tripetala* and *Paulownia tomentosa* are both cut back in spring; *Rhamnus frangula* 'Asplenifolia' and rosemary willow fill a wet, difficult corner. RIGHT Robust foliage of *Rheum australe* hugs the granite foundation wall.

Washington. This *Euphorbia* is marked by warm yellow tones in its bracts and flowers, and the rhubarb's bold texture with its fine foliage make for a nice contrast.

In between these two clumps grow other shrubs valuable for their leaves. *Salix fargesii* is most impressive, a multistemmed, medium-sized shrub with magnolia-like green leaves that emerge with reddish tones that extend into its petioles. Its winter buds are purplish red, and its wood waxy to the touch. Also impressive is *Gleditsia triacanthos* var. *inermis* 'Elegantissima', an old-fashioned honeylocust cultivar with fine foliage that I treat as a cutback shrub.

Gradually, I collected plants that felt right in the border, including another stunning rhubarb, *Rheum palmatum* 'Red Herald', the deepest red variety I

grow. It has dark red stems and deeply lobed dark green leaves, particularly conspicuous in cool spring weather, when its emergent red buds begin to extend. I'd seen it at Abbey Dore Court in England (now closed to the public), and when I found it for sale at the nursery at Heronswood, I knew I had to have it, even though it was priced at more than I'd ever spent for a single perennial. It's been worth it, as I've divided the original plant into a generous mass. Snakeroot (*Actaea racemosa*), planted behind it, is the perfect companion, and one of the few flowering plants I've let in. Its white candelabras freshen the border in July and August.

Two large fine-textured shrubs arch over the snakeroot and rhubarbs, anchoring the west end: rosemary willow (*Salix elaeagnos* subsp. *angustifolia*) and fernleaf buckthorn (*Rhamnus frangula* 'Asplenifolia'). The leaves of the former are silvery on the backsides and especially showy during dry spells, and when a breeze turns them up. The buckthorn's leaves are lacy, and

LEFT I transported both *Euphorbia palustris* 'Walenburg's Glorie' and a dark-leaved form of *Rodgersia podophylla* back from Heronswood Garden in carry-on luggage. **RIGHT** *Rheum palmatum* 'Red Herald'

the entirety of the shrub darker in appearance, contrasting with the willow. I chose fernleaf buckthorn not just for its foliage, but because it doesn't run or set seed and become invasive like some. It is a little-used plant, but one that draws lots of attention.

The two ends of the garden came together fairly easily, but the middle section took more time to get right. The middle portion of the wall was in poor shape, and I decided to mask it with two similar small-leaved climbing hydrangeas, *Hydrangea anomala* subsp. *petiolaris* 'Tiliifolia' and *H. anomala* subsp. *petiolaris* 'Jane Platt'. These came along slowly, as is their reputation, but their smaller leaves are more in scale with the wall and its surroundings. Less successful have been a variety of shrubs that have come and gone, either not liking the conditions or not quite the effect I wanted. Thorny devil's walking stick (*Aralia spinosa*), cutleaf elderberry (*Sambucus nigra* 'Laciniata'), *Salix magnifica*, and princess tree (*Paulownia tomentosa*) as a cutback plant

have all played a role at one time or another, but it's looking like two other tropical-looking shrubs will prove the most successful and long term here (I won't say permanent). One is umbrella tree (*Magnolia tripetala*), which I grow as a cutback shrub, with its large, ovate, shiny green leaves that sport silvery undersides. The other is a form of *Salix sacchalinensis* with long-reaching horizontal branches and narrow, glossy green leaves. Princess tree also grew here for many years, and was thought to be lost to winter kill in many years until its dinner plate–sized leaves emerged sometime in July.

Thanks to a couple of large aspens and birches on the terrace above, the border below the wall is shaded from midday sun. This has encouraged the growth of some of the largest-leaved hardy perennials, especially *Rodgersia podophylla*, *Astilboides tabularis*, *Darmera peltata* and *D.* 'Nana', and *Rheum australe*. Each of these lush plants thrives in the deep, rich soil and overhead

LEFT Climbing hydrangea runs along the top of the wall, while rodgersias, *Astilboides,* and *Darmera* thrive in moist soil at its base. RIGHT A jungle effect from cutback willows and magnolias among rodgersias and rhubarbs.

shade, and makes plausible the intended effect of this as a semi-tropical jungle.

These four gardens—the Long Border, New Meadow, the Cove, and Silver and Gold—may read as though they are four distinct ideas. In reality, they flow effortlessly into one another, tied together by long, sweeping edges cut into the lawn, each expressing its own character but viewed in progression. The generous size of the lawn and the length of the Long Border encourages lingering among the foliage; the massing of *Stephanandra* with its sweeping curve at the end of the border, together with a slight rise in elevation, invites movement forward but withholds the view of the meadow. The New Meadow is compact enough that its dusky beauty is easily inspected from the lawn. Although Silver and Gold is withheld from the view, there are hints that something lies ahead, and there is often a rush of adrenaline in visitors once it is apprehended. It is a short garden journey, rewarded with an array of experiences and emotions.

fruits and vegetables

One of the most gratifying, if time-consuming, aspects of the garden is producing a long season of vegetables, herbs, and fruits.

―――――――――――――

Vegetables are closely seeded in rows, allowing space for a hand-held cultivator, or occasionally interplanted with shorter-season crops such as lettuce or radicchio.

There is something fresh to use in the kitchen from late May into December, and although I don't put up a lot of food, there is tomato sauce, pesto, green beans, and applesauce in the freezer for winter meals, along with shallots, onions, and dried herbs to reach for in the pantry. Farming taught me how to produce crops early and in succession, and I still do that now, as space and time allow. I do not try to grow everything possible, but mainly concentrate on vegetables that are at their best harvested at the point of perfection, and just before being consumed, such as lettuces and salad greens, asparagus, and tomatoes. We're lucky to live near other first-class vegetable growers, and can visit local farm stands and farmers' markets for sweet corn, squashes, strawberries, and specialty crops.

I count myself lucky that I learned how to grow vegetables and don't mind the work involved. I get enormous satisfaction out of bringing produce in from the garden to enjoy and share with friends. I also have the usual complaints—crop failures due to adverse weather conditions and losses due to animal foraging. These come in cycles, and I try to tell myself they're a fact of life. Deer are not much of a problem in the Vegetable Garden or with fruit crops, but instead it is the small mammals—the woodchucks, chipmunks, and red squirrels—that do a surprising amount of damage in some years.

Most garden design manuals will instruct you to situate the vegetable garden out of sight, toward the rear of the property. I turned that convention on its head and placed the Vegetable Garden front and center, giving it prominence

The Vegetable and Flower Gardens line up into one composition, equally beautiful and productive in their own way. Pear trees frame the east end and apple trees the west. **OPPOSITE** Noble's Market Garden was short-lived, but the sign remains, as does my love of growing fresh fruits and vegetables.

almost equaling the Flower Garden. It was the only place on the property that made sense, and it felt right to me for the productive part of the garden to share equal billing with the ornamental part. I've made an effort to relate the Vegetable Garden to the Flower Garden by giving it similar proportions, linking the main path in the former to the central path in the latter. Corner posts holding up a wire fence give it some structure, and a couple of clematis also help tie it to the Flower Garden. It is a bountiful garden, with raised beds for vegetables, a long bed for herbs, rows of asparagus and raspberries, and a holding bed for nursery plants.

I focus mainly on crops that do well in our soil, and are productive over a long period of time: salad greens, carrots, beets, onions, shallots and leeks, pole beans and bush beans, Tuscan kale, tomatoes, eggplants, peppers, and annual herbs like flat leaf parsley and basil. I have experimented with crops that require more skill or technique to extend their season, but have largely foregone those recently. For many years, I purchased seeds from the Cook's Garden, then owned by Shep and Ellen Ogden. The Ogdens traveled in Europe searching out seeds of the most flavorful crops that grew well in American gardens, and it helped that they were cooks themselves and cared about flavor and freshness. It was a tremendous loss when they sold the company, when knowledge and passion were exchanged for corporate marketing. Today, Renee's Garden is my go-to supplier for the most flavorful crops of lettuces and salad greens, and Johnny's Selected Seeds and High Mowing Organic Seeds for much of the rest. Transplants come from three excellent organic growers.

I'm typically able to work the soil in the Vegetable Garden by the third week of April. There are six beds about 14 feet long by 4½ feet wide, with narrow paths between them. These beds are slightly raised above the compacted paths. They receive annual replenishment with compost, which I gently work into the top couple of inches of soil with a garden fork and a rake. I loosen the soil to the depth of the fork and use it or an iron rake to break up and crumble the top couple of inches, finishing it off with a light top dressing of North Country Organics Pro-Start, with its blend of micronutrients.

The rows are seeded closely, allowing width enough for a hand-held cultivator. Once the seeds are sown, a covering of spun polyester is placed on top of the soil to hasten germination and retain soil moisture; the spun polyester also keeps flea beetles off early-season targets like arugula.

The first spring sowing is lettuces, mesclun, and spinach, with Swiss chard and beets for early greens. I order a number of different lettuces and mesclun mixes, and sow short rows for a variety of tastes, colors, and textures. 'Bloomsdale Long Standing' and 'Tyee' are old standbys for spring spinach.

I can usually begin cutting lettuce by the third week of May, and start thinning chard and beet greens 10 days after that. By mid-May, I will have made a second sowing of lettuces and greens, and once those are in full production, the first sowings can be pulled out and room made for later sowings of carrots, beets, or green beans.

Shallots, leeks, and onion sets or starts get planted in mid-May; garlic is planted in fall. Members of the allium family do extremely well in our soil, and I now give them more space than I did in earlier years. It took me a long time to appreciate the difference in flavor that something like a homegrown onion can have. I like small shallots for vinaigrette and larger Dutch shallots to sauté or fry as a garnish. They bulk up fast, and the tops die back for a late August harvest. I lift the bulbs, let them dry on racks, and store them in baskets in a cool room in the house. Onions come along a little after that, and I pull them and let them dry on the ground where they grew. Once they have hardened off, with

Salad crops are sown as soon as the ground can be worked, usually the third week of April. Radicchio, leeks, carrots, and 'Sungold' tomatoes are mainstays of the Vegetable Garden.

their stems withered and necks tightened up, I gather them onto trays and store them in a dark corner in the ell, and finally bring them inside the house in loose bags. 'Patterson' is a long-lasting variety that keeps well. I allow leeks to stay in the ground longer—in milder climates they can be mulched and harvested throughout winter, but here I pull them just before the ground freezes and place them in buckets with moist sand in a cool, dark room, and use them copiously through the winter holidays.

I've cut back on some of the novelty crops, such as fava beans and artichokes. Both of these can be delicious fresh from the garden, but they take up too much room for the produce they yield, and we don't have the long, mild spring they favor. As much as I enjoyed the challenge of growing them, neither is ideally suited to growing here.

I set out transplants of tomatoes, eggplant, and peppers in late May. I watch the weather forecast, and my schedule, more than the phases of the moon. I transplant only a few of each, raised by nearby growers, rather than starting them myself. I've winnowed tomatoes down to just three or four varieties: 'Sungold' for its combination of sweet and acid in a cherry tomato that often starts bearing in mid-July; 'Hungarian Heart', an heirloom variety that I find tastier and more reliable than 'Brandywine'; 'San Marzano' for pasta sauce; and some years either 'Jet Star' or 'Morton', because they were two of the most popular tomatoes I grew as a market gardener.

Eggplants and peppers are grown for fresh picking and go straight on the grill for a vegetable-based August meal. They can be tricky, with mostly good years and some off years. 'Orient Express', with its long, slender fruits, is the most reliable and productive eggplant in my garden. It bears a couple of weeks earlier than the Italian types, with the first fruits harvested in mid-July and continuing well into September if the weather cooperates. For peppers, poblanos are versatile and great on the grill too, with extras sliced into strips and frozen to fry with potatoes and onions in winter. 'Carmen' is a sweet Italian bull's horn pepper best for frying and grilling.

Green beans have many uses and I grow a number of varieties—too many in some years—but the excess goes into the freezer for winter soups. I typically grow both pole beans and bush beans, the bush beans in full or half rows as succession crops, with the first sowing once the soil has warmed in mid-May. I try to keep the harvest coming at a reasonable pace and not overwhelm the kitchen with too many at one time. I plant pole beans a little later, when the soil and nighttime temperatures are reliably warm; I usually plant two or three varieties and harvest handfuls at a time.

'Provider' is a mainstay bush bean, germinating in cool soils, an adaptable grower and early bearer. It is not the most flavorful, but is dependable and eagerly consumed. 'Orient' has more flavor and is best served as a whole bean. 'Tavera' is also easy to grow, its medium-sized beans for haricot verts to be savored in salads, or as a meal unto themselves with a shallot vinaigrette. Italian flat pod beans have a meatier flavor that makes for a much-anticipated stew of Roma beans, onions, and tomatoes, seasoned with thyme and oregano.

Pole beans yield some of the richest flavors and can bear over a more extended period than bush varieties. For many years I grew 'Trionfo Violetto', a purple Italian variety whose color makes it easy to pick and fades to a hearty green when blanched. I find 'Northeaster', a vigorous, early-maturing, mostly stringless Roma-type pole bean, worth the effort for its rich, nutty flavor. 'Fortex' is a long pole bean that also packs flavor.

All of these are worth growing because when they're ready to pick, they determine the menu, and their absolute freshness means they have more flavor even than those purchased at a farmers' market.

Lacinato kale is a long-season crop, harvested young for Bagna Cauda, and in fall to be chopped and dressed for late-season salads.

Leafy greens like Swiss chard have a longer season. It is seeded early and rapidly gains in size, its young leaves and crunchy white stalks are excellent blanched and dressed with olive oil and lemon. It also provides the impetus for bringing out the fondue pot and preparing a meal of Bagna Cauda, where garlic and anchovy are gently wilted in olive oil and become the "warm bath" into which the leaves and stalks of Swiss chard are dipped, along with many young gatherings from the garden: baby green beans, zucchini, Tuscan kale, lettuce, baby carrots, and cherry tomatoes. Each is dipped into the oil bath and a meal is made from whatever vegetables are on hand, along with rustic bread. Once the vegetables are consumed, a couple of fresh eggs are stirred into the

LEFT An old tree bears modest crops of tart apples good for sauce and cider. MIDDLE Apple blossoms perfume the May air while the sound of bees working the flowers generates a buzz that says spring. RIGHT A minimal spraying program for the apples keeps most diseases at bay and encourages a healthy harvest.

remaining mixture, and it's all finished off with spoonsful of anchovy-infused scrambled eggs on crusts of bread.

Tuscan kale, or lacinato kale, is also sown early or transplanted from starts. It makes for a delicious salad, its tender, leafy greens stripped away from the stalks and shredded into a bowl, mixed with bits of salmon and pistachios, and dressed with olive oil and vinegar.

When the first sowing of beets is thinned, their greens, with the beets just starting to form, are steamed or briefly boiled and then dressed with oil, lemon juice, or vinegar. Later on, when I harvest the beets, these are boiled or roasted, and chopped into a salad with crumbled blue cheese and toasted pecans or walnuts.

Almost every one of these vegetables reaches its maximum potential when paired with fresh herbs from the garden: basil for tomatoes, oregano for eggplant, mint for peas, shallots and thyme for green beans. I dedicate one row to herbs, including a hardy French tarragon, chives, Greek oregano, lovage, lemon balm, two kinds of mint, English and French thyme, winter savory, marjoram, rosemary, and sage. I seed basil and parsley annually in the beds.

Two rows of 'Jersey Knight' asparagus yield a month's worth of meals. This variety has large spears and performs exceptionally well in heavier soils. Excess production goes to friends, who report it is the sweetest, most succulent asparagus they've ever tasted. In a row parallel to the asparagus grow four varieties of raspberries: 'Boyne' is an early bearer, followed by 'Latham', and then 'Polana' and 'Jaclyn', which are everbearing types that can start in August and go until frost.

Three pear trees dominate the lawn east of the Vegetable Garden, each a distinct cultivar. 'Ure' bears small, juicy fruit that can be eaten right off the tree, while 'Luscious' has superior flavor, but needs to be chilled before the fruits fully ripen to attain its best flavor. Those of 'Seckel' are small, but their spicy flavor is the essence of pear juice. 'Luscious', which also has beautiful glossy green foliage, usually bears well, while the others have years of heavier or lighter production. In bountiful years, the trees are stripped of their fruits, and these are pressed into pear juice, some of it turned into cider.

The apple orchard on the west end of the garden can also produce a bountiful crop of apples for eating, storing, and cider making. The original planting of 'McIntosh' trees that had been dug from an active orchard and planted primarily for screening usually needs an assist to bear good quality apples—three sprayings in spring and early summer with a blended organic-synthetic fungicide. Much more reliable are the two 'Liberty' trees and a large 'Cortland', whose fruits can be stored in the cool cellar to use for applesauce in fall and early winter.

None of the apples goes to waste. Every other Saturday during apple season, a group of neighbors gets together and picks all the apples from the older trees and remnant orchards on Bragg Hill. It's not uncommon for a group of 20 kids and adults to strip the trees, the younger ones climbing the trees and shaking limbs and the rest of us picking up the drops. The older trees with the tarter apples make the best cider. The entire crop is picked by mid-October and pressed into Deer Run Cider, made by the neighbors on Bragg Hill.

The McKenzies and those who farmed here before them had to rely on what the farm could produce as their main source of food year-round. We're fortunate in that we can bring in something from the garden from May into November, and reach for onions, shallots, herbs, or frozen tomatoes the rest of the year. We are dependent on the garden in a different way, but it still feels like an essential part of living here.

field and forest

Encouraging wildlife is an important goal for the garden. Hayfields and pastureland were part of the working farm, and these open grasslands now support birds and other creatures.

Shrubs like staghorn sumac and hills of snow hydrangea benefit a variety of wildlife.

Native shrubs and plants from its agricultural past help sustain wildlife and integrate the garden into the open lands. Native wildflowers in the garden and meadow also encourage pollinators, and annual late summer mowing is timed to benefit birds that nest in the fields. The second-growth woodland here will eventually become a functioning part of the maple-beech-birch forest type that dominates the region. The garden is but one part of a larger ecosystem and cycle of life.

garden meets landscape

At times, I become acutely aware of how the garden sits within the larger living world. Sometimes it is during a midsummer squall, when every tree sways as if it's about to snap, or it may be when the garden is buried under drifting snow, its lines blurred. This feeling is, perhaps, most acute when garden, field, and woods are permeated by the sound of migratory birds returning for their brief season in the north. The inhabitants of the neighboring meadows and forests treat the garden as an additional source of food and shelter. It is a part of a dynamic, living landscape.

The garden sits adjacent to open meadows, mixed deciduous forest, and woodland edges. It functions somewhat like a woodland edge itself, with part of it bordering on meadow and part giving way to forest. It has expanses of sun-loving meadow and prairie plants, as well as shaded areas where woodland perennials predominate. Parts of the garden that began open and sunny are becoming more like woods as trees and shrubs planted 20 years ago gain in stature and their canopies knit together.

The garden is a relatively small part of our property: our 10 acres of woods is but a fragment of 4 million forested acres in Vermont, and the 10-acre field is just one parcel of nearly 400 contiguous acres of open fields along Bragg Hill. The combination of woods and meadows and the intersection where mown fields meet forest creates a habitat for diverse vegetation and animal life.

My aim is for the garden to be a functioning part of the surrounding world, and for it to participate as best it can in the natural processes going on around it, in both the agricultural and forested landscapes. I want the garden to help support the wildlife and organisms that live their lives in the fields and woods about. As populations of insects, bees, butterflies, moths, and the birds that feed on them fluctuate and decline, it has become more critical to take into account ways in which the garden can help sustain these life forms. Part of that

The return of migratory birds for their brief summer stay is an acute reminder for me that the garden is part of the larger living world.

is to provide habitats and plants compatible with those found in the region, and another part is to be a good steward of the fields, woods, and edges.

I've come to see that the garden can fulfill different functions—not just my passion for working with plants to create garden beauty, but for it to be a source of nourishment and habitat for wildlife. I first came to this realization at the beginning, pulling blackberries that threatened to engulf the house in preparation for our repair of its crumbled foundations. It took a while to pull out the brambles, and in doing so, I realized I was hearing less and less birdsong, and that my efforts to clean up were removing nesting and perching habitat. There was no question the brambles had to go, but this introduced me to the idea that the garden would need to include shrubs and other dense growth at the edges of the field to bring the birds back. Now, instead of being overrun by brambles, we look out the kitchen window through stems of the red berries of winterberry (*Ilex verticillata*) and across mounds of smooth hydrangeas, backed by native

LEFT Shrubs like pagoda dogwood are planted or encouraged at the edge of fields to bring birds closer to the garden. RIGHT *Hydrangea arborescens* 'White Dome' is one of the best of the native hydrangeas for attracting all sorts of bees and hoverflies in the garden.

dogwoods and sumac. It's a picture as pleasant to view in winter as in summer, and one that brings cedar waxwings close to feed.

This early revelation of how the loss of habitat for birds was also a loss for us encouraged me to think about how the garden is part of the ecology of the place, and to examine what consequences my gardenmaking might have for the landscape around. This didn't result in my making butterfly or pollinator gardens, but rather asking how I could incorporate plants that have wildlife and conservation value into gardens that were largely driven by design. It encouraged me to pay attention to which plants, both native and introduced, might have the highest value for birds and pollinators, and to utilize those plants when they could succeed both functionally and aesthetically. There are plenty of lists of plants that attract pollinators and provide food and cover for

FAVORITE PLANTS FOR WILDLIFE

These plants succeed both functionally and aesthetically, and appeal to all manner of wild creatures:

TREES AND SHRUBS

Birch (*Betula* spp.)

Crabapple and apple (*Malus* spp.)

Cranberry (*Vaccinium macrocarpon*)

Dogwood (*Cornus* spp.)

Honeysuckle (*Lonicera* spp.)

Maple (*Acer* spp.)

Oak (*Quercus* spp.)

Pear (*Pyrus* spp.)

Poplar (*Populus* spp.)

Smooth, panicle, and climbing hydrangea (*Hydrangea arborescens, H. paniculata, H. anomala* subsp. *petiolaris*)

Spruce (*Picea* spp.)

Stephanandra incisa and *S. incisa* 'Crispa'

Viburnum spp.

Willow (*Salix* spp.)

Winterberry (*Ilex verticillata*)

PERENNIALS

Allium 'Millenium'

Anise hyssop (*Agastache* spp.)

Aster (*Aster* spp. and *Symphyotrichum* spp.)

Bee balm (*Monarda* spp.)

Cardinal flower (*Lobelia cardinalis*)

Fireweed (*Epilobium angustifolium*)

Garden phlox (*Phlox maculata* cvs. and *P. paniculata* cvs.)

Gentian (*Gentiana* spp.)

Goldenrod (*Solidago* spp.)

Helenium spp.

Joe Pye weed (*Eupatorium* spp.)

Milkweed (*Asclepias* spp.)

Nepeta (*Nepeta* spp.)

Perennial sunflower (*Helianthus* spp.)

Snakeroot (*Actaea racemosa, A. simplex*)

Thyme (*Thymus* spp.)

Veronicastrum virginicum and *V. sibiricum*

birds and mast for wildlife, but it is especially rewarding to discover when a plant performs well both environmentally and ornamentally. Such an observation came when I noticed how much more bee and syrphid fly activity there was on the selection *Hydrangea arborescens* 'White Dome' than on the stand of the straight species nearby. 'White Dome' may not behave this way in every garden, or even every year in my garden, but the lesson for me is that experimentation and diversity in plants is to be encouraged, and that close observation and informed judgment is part of what it means to cultivate one's own garden.

In *Bringing Nature Home*, wildlife ecologist Doug Tallamy stresses the importance of selecting native plants that support insects that feed on them, insects that in turn are the basis for the birds, reptiles, amphibians, and mammals that ingest them as a food source. Tallamy encourages the use of plants that have evolved together with insects and share their physiological adaptations and specializations. The oaks, cherries, willows, poplars, birches, and maples that support native butterfly and moth species all grow in the surrounding woods, and some have found a place in the garden and its margins. Asters, goldenrods, Joe Pye weed, sedges, violets, phloxes, and milkweeds grow abundantly along nearby edges of fields and woods, and are also woven into the garden.

I've planted a broader and more diverse range of plants in the garden than might grow naturally along the field edges, to provide nectar and other sustaining nourishment. Plants initially selected for ornamental value can also play a role in attracting insects and birds. Floriferous garden plants like roses, hydrangeas, and gentians are made more interesting and useful by the pollinators that are attracted to them. Larval plants for Lepidoptera include many less showy but still valuable garden plants such as turtlehead, pussytoes, little bluestem, violet, parsley, and lovage. Field-growing nectar plants like milkweed and New England aster provide sustenance too, as do more traditional garden plants such as *Liatris*, sweet pepperbush (*Clethra alnifolia*), Joe Pye weed, and purple coneflowers.

My interest in extending the season of bloom here and expanding the period of garden interest has had the unintended result of providing food sources for wildlife over a longer period than would normally be found in the surrounding fields. *Helleborus thibetanus* attracts early pollinators as the snow

At the garden's margins, the richness of habitats and plants is amplified, including native dogwoods and hydrangeas, willows, and spruces.

melts, and *Symphyotrichum oblongifolium* 'October Skies' supplies pollen and nectar weeks after the native New England asters have faded. Seed heads of coneflower and cardoon supplement the seeds of gray birch and thistle that goldfinch feed on late in the season.

EDGES

At the garden's margins, I've tried to amplify the richness of habitat and plants found at the intersection of sun and shade, meadow and forest. A single red-stemmed dogwood (*Cornus sericea*), for example, was planted in exactly the conditions in which it thrives in the wild—organic, consistently moist soil in full sun—and it's grown 20 feet wide and 12 feet tall. Its red stems are an ornamental feature in winter, but its real value is as a boundary marker for the garden, as well as cover for nesting birds. In summer, its healthy green foliage remains trouble-free. Its white flowers may be lackluster, but they mature into bluish drupes that the birds strip in fall. On higher, drier adjacent ground, I transplanted staghorn sumac (*Rhus typhina*), also as an edge to the garden, and have watched it spread toward even drier ground. Its gnarly stems offer shelter to ruffed grouse and turkeys, which feed on its red hairy fruits.

Farther along, at the boundary of meadow and lawn, I planted four rosemary willows (*Salix elaeagnos* subsp. *angustifolia*) from rooted cuttings. These willows grow in a drier spot than the dogwood, and have remained fairly low—noticeably smaller than the specimen growing in the Long Border, where the soil is moister and there is no competition from grasses. There are paths worn under the willows from wildlife moving between the meadow and garden. Male bobolinks perch on their upper branches and chirp "spink, spank, spink" between courting display flights. Bees seek out this willow in spring, along with many of the others interspersed throughout, these being among the earliest nectar-bearing plants to bloom.

Next to the willows, high up in a crotch in the Lombardy poplars on the field's edge, a female Baltimore oriole weaves a hanging nest of bark and grass. Although the nest is sometimes difficult to make out, it draws attention when the males dart from the branches to snag passing insects. The Flower Garden's upright cedars offer protected nesting sites for catbirds, where the dense cover allows them to hunt for beetles and caterpillars feeding on the nearby flowering perennials.

The line of Norway spruce on the property edge is cover for nesting pairs of mourning doves, who forage by day on seeds and grains found in the open

Staghorn sumac has colonized a drier part of the margins.

fields and garden, returning to their roost at dusk. These birds enliven gray autumn skies when they return by the dozen in the evening and dive into the trees' upper branches.

The assortment of shrubs, trees, and perennials in Silver and Gold is actually a cultivated edge to a large, scrubby understory of bottlebrush buckeye, native dogwoods, willows, elderberries, and smooth hydrangeas. It is where I sometimes deposit leaf litter from the rest of the garden; this slow accumulation of organic matter has encouraged the growth of mayapples, ferns, and a few spring ephemerals.

This dense undergrowth is perfect habitat for ruffed grouse, and a female might scratch out a nest on the ground while remaining camouflaged by the surrounding vegetation. Grouse can feed on the shrubs, ferns, and perennials as well as nearby buds of poplars and birches. More than once, both grouse and I have startled one another face to beak. And aside from grouse, one fall, a young bear took refuge in a stone wall in this thicket to be near the ripening apples in the orchard.

I established a more shaded edge under one of the apple trees with Christmas ferns and marginal wood ferns, along with sedges and some trilliums, covering the ground in between with leaves. The trilliums increased rapidly, and the ferns found it fertile ground. The longer this planting settles in, the more it takes on the character of a native woodland edge, and the less it reads as a garden. Selective weeding and a restrained hand in adding plants keep it from appearing too garden-like.

Farther along the road are remnants of former gardens with decades-old roses and hydrangeas. Coltsfoot blooms in the roadside ditches. Marsh marigold sits in snowmelt, and is followed by trilliums, columbines, and *Tiarella*, then baneberry and Jack-in-the-pulpit. The roadside goes quiet in summer, with the subtle charms of ferns appearing under the foliage of moosewood and

LEFT Wildlife wears paths under rosemary willow, and in spring, bobolinks and other edge- and field-nesting birds move between the willows and poplars. MIDDLE Neighbors pick the apples to press for cider. The remainder feed deer and turkeys. RIGHT Dense undergrowth encourages native woodland plants to grow and spread, as well as providing a home for ground-nesting birds.

purple-flowered thimbleberry. In late summer, the asters begin a long show, with tall purplestem aster (*Symphyotrichum puniceum*) growing alongside sensitive fern in wet ditches, and hairy white oldfield aster (*S. pilosum*) on drier ledges. As the leaves begin to fall from the trees, the ferns stand out again, along with tawny leaves of beeches that catch the lowering light.

These edge habitats also harbor animals that feed on the garden, sometimes to its detriment. The drops from the apple trees provide food for wild turkey and deer as they bulk up for winter; bears have occasionally

LEFT The field is mown after the departure of the bobolinks, often not until September. RIGHT Along the field's richer edge, where daffodils bloom in spring, mowing is delayed until October to give milkweed a chance to increase.

stripped the pears from their trees just as they turned ripe. Crows, red squirrels, and chipmunks—the latter two forest dwellers that have adapted to civilization—are also attracted by the abundance of food types available from the garden. Seeds, nuts, fruits, and the stems of trees are all part of their diet, and they gather seeds and cones and store them for winter. In autumn, the squirrels will sever green cones from spruce trees, piling them into mounds to return to when winter comes. In spring, clusters of a dozen red currant or birch seeds that chipmunks buried but forgot will sprout. Red squirrels knock the pears from their trees, and may take only a few bites before moving on to the next. Meadow voles and field mice are as at home in the fields as in the garden, and will burrow under the snow to feast on the green stems of sedges and grasses; worst of all, they will girdle the juvenile bark of shrubs and fruit trees.

Woodchucks, cousins to the squirrel, have built burrows in the meadows and stone walls of the farm for as long as the land has been cleared. They will feed on the succulent new growth of grasses, and are especially attracted to newly sprouted sowings of beets, carrots, and salad crops. But the real damage they perpetrate is in undermining the foundation walls of the house and barn.

Deer can be a garden's demise, and they're frequent visitors here, but so far this garden has survived. I have had to learn their habits and employ a variety of tactics to deter them. Evergreen cedars and junipers, which are so essential to the garden's design, are covered with netting and wire fencing just before Thanksgiving; the netting is left on until the deer start feeding on fresh meadow grass in spring. In summer and early fall, when they begin browsing on the buds and foliage of native shrubs and perennials, an application of a deer repellent is often enough to deter them. It is only because there are still hunters in this suburbanizing landscape that the garden can survive.

MEADOWS

The remnant hayfields still support forage and hay crops such as timothy, smooth bromegrass, reed canary grass, and white and red clover. The edge closest to the farmyard has richer soil where grass grows thicker and taller, with only a few forbs interspersed. Milkweed, food source for the monarch butterfly, is more plentiful in the richer soil at the edge of the meadow, where narcissus blooms in spring. We don't mow this area until very late in the season, if at all, and the milkweed is gradually increasing its footprint. Goldenrod is thick along stone walls at the edge of the woods.

The gently sloping 10-acre meadow is a matrix of field grasses, native and introduced wildflowers and weeds, whose extended period of bloom creates an impressionist painting. In the drier portions, a mixture of yellow and white meadow flowers blooms from spring through summer. These include dandelion, golden Alexander, June daisy, buttercup, blue-eyed grass, wild strawberry, red sorrel, yarrow, hawkweed, wild carrot, black-eyed Susan, and goldenrod. In the wet seams, sensitive fern and horsetail have colonized.

This meadow bears out the theory behind shortgrass matrix plantings for constructed meadows, in that the soil's low fertility keeps the grasses in check, and allows a selection of forbs to persist. The field hasn't been limed or fertilized or otherwise improved in 40 years. The grasses outcompete the forbs at

A matrix of field grasses, native and introduced wildflowers, and ferns, growing in their preferred niches in the field, produce an extended period of bloom and an impressionist painting over the course of the growing season.

the richer garden edge, but the main flowering interest is in the lower fertility areas where forbs share the ground with the grasses that have persisted.

Woodcock and bobolink routinely nest in the meadow because we delay the annual mowing until after they've fledged. In more productive hayfields, local farmers now delay a first mowing on up to 40 percent of the field in order to leave habitat for ground-nesting birds. Enlightened property owners are reducing the size of lawns that are mowed every week, adding to potential nesting habitat. Our field is cut in late July or even as late as September, either for mulch hay, which is baled and removed, or brush-hogged and the grass left to decay. A fall hay crop is sometimes taken off. The fields must be mowed annually, otherwise white pines and birches would soon begin to reclaim the meadows as forestland.

The mating ritual of the woodcock high above the meadow is an early harbinger of spring. The field remains alive with bird activity until the bobolinks gather into flocks in late summer in anticipation of their long journey to the forests of Central and South America. Crows and ravens take possession of the field later in summer, attended by red-tailed hawks, kestrels, and turkey vultures. Fireflies, once a staple of early July nights, are in decline, and most unnervingly, are sometimes alight on Memorial Day rather than the Fourth of July.

WOODS

For a farm in the 1850s to produce 400 bushels of potatoes and 450 pounds of wool, there would have been sizeable acreage of open fields, and that openness would have continued into the early 1900s, when 100 milk cows on the farm supplied dairy products to Dartmouth College. Over the course of the 20th century, as production waned, fields reverted to trees, and long stone walls that would have separated sheep pasture were swallowed up by the woods.

An aerial view from the 1950s shows established hardwoods and hemlocks growing along the stream corridor at the far edge of the property. The wooded area we now see from the house is early successional regrowth, primarily white pine, with a few stands of poplar and red maple in wet swales. In six decades, the pines have reached 70 feet in height and are now starting to fail where their feet are wet.

Buckthorn, an invasive scourge in New England, has unfortunately invaded much of the understory beneath the pines. Some years ago, in a cooperative effort with neighbors who were clearing white pine from their land, we did a selective thinning of some taller pines that were impeding our view. In that swath, there is now a succession of hardwoods coming along, primarily beech.

As the land descends to the brook, the forest type changes noticeably: oak, beech, maple, hemlock, and yellow birch predominate on the slopes. The woodland floor is carpeted with Christmas fern and marginal wood fern, with a few sedges, wake robin (*Trillium erectum*), and false lily-of-the-valley growing about. Woodland edges reveal baneberry, false Solomon's seal, Jack-in-the-pulpit, a variety of violets, and *Tiarella*.

Closer to the brook, where the air is moister, maidenhair ferns grow out of seeps, and royal ferns grow in pockets of duff. A dense grove of hemlock offers winter protection for deer. The woods are also home to birds we hear but seldom see. Hermit thrush, veery, and vireo are the sound of Vermont summer

As the land descends to the brook, the forest type changes from white pine to a mix of hardwoods. The woodland floor is carpeted by Christmas fern, marginal wood fern, sedges, and trilliums.

evenings, and a barred owl hoots from time to time. A more chilling sound that emanates from the woods is the eastern coyote's yipping and howls when packs reunite at night.

Observing the seasonal rhythms in the fields and the slower pace of the forest sometimes leads me to reflect about where the garden is headed, and how it will change as I slow the pace of its development and become less able to provide all the care it needs. In the sunny Flower Garden, Rock Garden, and Stable, I am concentrating on long-lived plants that will grow together as a community. The main challenge is to keep the taller and lusher native herbaceous perennials from swamping lower-growing, more cultivated plants in the Flower Garden. Some of the intricate combinations that attract the most attention are giving way to denser flowering shrubs.

Shadier beds are a bit easier to maintain as a fairly stable system, although they do require keeping a vigilant eye open to rein in the more vigorous colonizers and cull self-sowing native lady fern. Any garden, sun or shade, requires grooming to keep it looking fresh and cared for.

A lesson in where the garden may head came my way when, after removing the remnants of the sour cherry tree from the kitchen bank, I decided to wait a while before replacing it, hoping for some inspiration or for the right plant to make itself known. I made a list, and looked at plants in nurseries, but mostly paid attention to other matters. Eventually I noticed that a seedling pagoda dogwood (*Cornus alternifolia*) was growing out of the vinca on the slope, probably via seed dropped by a bird. While pagoda dogwood isn't known to be particularly long-lived and doesn't transplant well, my experience at The Fells told me that this native tree can persist for decades, and with its horizontal branching and reddish purple fruits, makes a pleasing statement in the garden. So far it has demanded little of me. I will watch it grow and hope this gift from the birds will prosper and bring years of sustenance and joy.

Who knows where the garden may head?
I plan to watch it grow and be thankful for this gift.

ACKNOWLEDGMENTS

This book was conceived in a time of hopefulness, when it looked like America could somehow face the challenges posed by a rapidly changing climate and diversifying social fabric with a more resilient vision of the future. But it was written during the dark days of retreat and accusation. I write from a place of optimism and hope for a future where others who do not wish to inhabit a world of relentless bad news might find value in an engagement with plants, the beauty of gardens, and sharing the sustenance they provide.

My garden has been made in the context of regular exposure to some extraordinary personal gardens and the people who made them. Many of them have helped stretch my vision of what a garden is and the plants that could thrive in it. I thank them for their gifts of confidence, the plants they have shared, and their dedication to making and preserving gardens: Antonia Adezio, Marco Polo Stufano, George Schoellkopf, Frank Cabot, Tom Armstrong, Nancy Goodwin, John Fairey, Fergus Garrett, Lucy Hardiman, and Don Avery to name just a few.

Numerous colleagues at the Garden Conservancy and its preservation projects encouraged this self-taught gardener to roll up his sleeves to help others preserve their exceptional creations: Diane Botnick, Laura Palmer, Claire Sawyers, John Trexler, Dick Lighty, Patti McGee, Joe and Anne McCann, Colin Cabot, Page Dickey, Russ Beatty, Dick Turner, Amy Graham, Ann Loeffler, Gusta Teach, Bobbie Dolp, and Rosina McIvor.

Gardens I came to know through my association with the Garden Conservancy and the many volunteer activists dedicated to saving and sharing them with the public are a part of my garden journey, including the gardens of Alcatraz, the Ruth Bancroft Garden, and Western Hills in California; the Chase Garden in Washington; the Lord and Schryver Conservancy, Jane Platt garden, Elk Rock, and Ernie and Marietta O'Byrne's garden in Oregon; Peckerwood Garden in Texas; Longue Vue House and Garden in Louisiana; the Elizabeth Lawrence Garden and Montrose in North Carolina; the Pearl Fryar Garden in South Carolina; Greenwood Gardens and Meadowburn Farm in New Jersey; Long House Reserve, Madoo, Rocky Hills, and Stonecrop in New York; The Fells in New Hampshire; and Shelburne Farms in Vermont.

The opportunities afforded me by the stewards of Cornish Colony gardens have been invaluable. My sincere thanks go to Charles and Joan Platt, Max Blumberg and Eduardo Araújo, Bob Gordon and Marjorie Mann, and the staff of the Saint-Gaudens National Historical Park.

I am grateful for the plant explorers, growers, and nursery owners who help make this a plant-rich garden: locally Cady's Falls Nursery, E.C. Browns' Nursery, Killdeer Farm and Perennial Pleasures Nursery in Vermont, as well as Edgewater Farm in New Hampshire; Broken Arrow Nursery in Connecticut; Opus (issima) in Rhode Island; and Garden Vision Epimediums in Massachusetts. A little farther afield: Plant Delights Nursery in North Carolina; Heronswood and Far Reaches Farm in Washington State; Joy Creek Nursery, Northwest Garden Nursery, Forestfarm, and Edelweiss Perennials in Oregon; Digging Dog Nursery in California; and the plant sales of many public gardens.

Closer to home, a number of people have shared advice, enthusiasm, their time, plants, and material objects to help improve my garden: Alan Saucier and Bill Flynn, Herb Ferris, Bill Murphy, Cal Felicetti, Margie Carpenter, Jude Powers, Gary and Mikayla Spaulding, Mike and Gertie Luce, Julie Welch, the staff of E.C. Browns' Nursery, and the many neighbors who walk by and offer words of encouragement.

The staff of the Norwich Historical Society and Dartmouth College Library guided me to photographs and archival information.

I am especially grateful to friends and colleagues who encouraged me to share my garden in print, especially those who helped me find a more concise route to the finish: Judith Tankard, Florence Fogelin, Penny McConnel, Diane Botnick, Laura Palmer, Tom Fischer, Andy Keys Pepper, and Jane Lincoln Taylor. Rita Cruise O'Brien helped move the project from idea to reality. They all earn the thanks of my patient mother Chickie Noble, who is eager to see this in print. Thanks go to Ivanna Folle and the Bogliasco Foundation for their support to help get the project off the ground.

I enjoy sharing the garden with all comers, but the greatest sharing has been with Jim and Sue. Friday is Sue's regular day in the garden, and at the end of a long day Jim would join us for a last look at what we had accomplished before sitting down to a meal together. This has been the greatest act of sharing in my life and I am forever grateful for all our Fridays together.

GARDENS TO VISIT

The following is a list of gardens that have been touchstones for me and reflect the scope and aspirations for my own garden.

*Private garden only open to the public occasionally.   **Open by appointment only.

north america

Abkhazi Garden
Victoria, British Columbia, Canada
abkhaziteahouse.com

Bellevue Botanical Garden
Bellevue, Washington, USA
bellevuebotanical.org

Bloedel Reserve
Bainbridge Island, Washington, USA
bloedelreserve.org

Chase Garden*
Orting, Washington, USA
chasegarden.org

Elizabeth Lawrence Garden
Charlotte, North Carolina, USA
winghavengardens.org/
elizabeth-lawrence-house-and-garden

The Fells Historic Estate & Gardens
Newbury, New Hampshire, USA
thefells.org

Gaiety Hollow
Salem, Oregon, USA
lordschryver.org

Greenwood Gardens
Short Hills, New Jersey, USA
greenwoodgardens.org

Heronswood
Kingston, Washington, USA
heronswoodgarden.org

Hollister House Garden
Washington, Connecticut, USA
hollisterhousegarden.org

Innisfree Garden
Millbrook, New York, USA
innisfreegarden.org

Jane Platt Garden*
Portland, Oregon, USA

Juniper Level Botanic Garden at Plant Delights Nursery*
Raleigh, North Carolina, USA
juniperlevelbotanicgarden.org

Landcraft Environments*
Mattituck, New York, USA
landcraftenvironment.com/lc_gardens.html

Les Jardins de Quatre-Vents * (open by reservation)
La Malbaie, Quebec, Canada
cepas.qc.ca/jardins-quatre-vents/
acheter-des-billets/

LongHouse Reserve
East Hampton, New York, USA
longhouse.org

Madoo
Sagaponack, New York, USA
madoo.org

Marietta and Ernie O'Byrne Garden**
Eugene, Oregon, USA
northwestgardennursery.com/our-garden/

Montrose**
Hillsborough, North Carolina, USA
montrosegarden.org

Native Plant Trust (Garden in the Woods)
Framingham, Massachusetts, USA
nativeplanttrust.org/visit/garden-woods/

Peckerwood Garden
Hempstead, Texas, USA
peckerwoodgarden.org

Ruth Bancroft Garden
Walnut Creek, California, USA
ruthbancroftgarden.org

Saint-Gaudens National Historical Park (Aspet)
Cornish, New Hampshire, USA
nps.gov/saga/index.htm

Sakonnet Garden*
Little Compton, Rhode Island, USA
sakonnetgarden.net

Stonecrop Gardens
Cold Spring, New York, USA
stonecrop.org

Untermeyer Gardens
Yonkers, New York, USA
untermyergardens.org

Wave Hill
The Bronx, New York, USA
wavehill.org

Western Hills Garden*
Occidental, California, USA
westernhillsgarden.com

europe

The Beth Chatto Gardens
Colchester, Essex, UK
bethchatto.co.uk

Dr. Jac. P. Thijssepark
Amstelveen, Netherlands
thijssepark.nl

Great Dixter House & Gardens
Northiam, Rye, East Sussex, UK
www.greatdixter.co.uk

Foerstergarten (Karl Foerster's garden)
Potsdam, Germany
denkmalschutz.de/denkmal/
Wohnhaus-und-Garten-Karl-Foerster.html

Jardin du Vasterival
Sainte-Marguerite-sur-Mer, France
vasterival.fr

Prospect Cottage (Derek Jarman's garden)*
Dungeness, Kent, UK

Tuinen Mien Ruys
Dedemsvaart, The Netherlands
tuinenmienruys.nl

Weihenstephaner Gärten (Sichtungsgarten
Weihenstephan)
Freising, Germany
hswt.de/weihenstephaner-gaerten.html

Villa Gamberaia
Settignano, Florence, Italy
villagamberaia.com/i-giardini/

Villa Serbelloni
Bellagio, Como, Italy
villaserbelloni.com

PHOTO CREDITS

All photos by the author, except for the following:

Courtesy of Dartmouth College Library, pages 26 upper right, middle left, and lower right, 29 upper left, upper right, middle right, and lower right, 31, and 80 upper right

Ellen McGowan Biddle Shipman papers, #1259; Division of Rare and Manuscript Collections, Cornell University Library, page 80 left

Courtesy members of the Platt Family, page 26 lower left and 80 lower right

James Tatum, page 52 left

INDEX